T0327434

Integrated Powertrains and their Control

Integrated Powertrains and their Control

Edited by

N D Vaughan

**Professional
Engineering
Publishing**

Published by Professional Engineering Publishing Limited,
Bury St Edmunds and London, UK.

First Published 2001

ISBN 978-1-86058-334-6

A CIP catalogue record for this book is available from the British Library.

Related Titles of Interest

Title	Editor/Author	ISBN
Advances in Vehicle Design	J Fenton	1 86058 181 1
Vehicle Systems Integration – The Way Ahead	Edited by A V Smith and C Hickman	1 86058 262 1
Automotive Engines and Powertrains	IMechE Seminar	1 86058 114 5
Integrated Powertrain Systems for a Better Environment	IMechE Conference	1 86058 224 9

For the full range of titles published by Professional Engineering Publishing contact:

Sales Department
Professional Engineering Publishing Limited
Northgate Avenue
Bury St Edmunds
Suffolk
IP32 6BW
UK

Tel: +44 (0)1284 724384
Fax: +44 (0)1284 718692
Website: www.pepublishing.com

Contents

Authors' Details

Details of contributing authors are listed below.

Chapter 1 – Introduction to Advances in Powertrain Technology
N S Jackson
Ricardo, UK

Chapter 2 – Control of an Integrated IVT Powertrain
S Murray
Powertrain Control Department, Torotrak (Development) Limited, Leyland, UK

Chapter 3 – Driveability Control of the $ZI^®$ Powertrain
A F A Serrarens
Faculty of Mechanical Engineering, Section Systems, and Control, Eindhoven University of Technology, The Netherlands

Chapter 4 – Performance of Integrated Engine-CVT Control, Considering Powertrain Loss and CVT Response Lag
T Kim and *H Kim*
School of Mechanical Engineering, Sungkyunkwan University, Suwon, Korea

Chapter 5 – Shifting Dynamics of Metal Pushing V-Belt – Rapid Speed Ratio Variations
G Carbone and *L Mangialardi*
Dipartimento di Progettazione e Produzione Industriale, Politecnico di Bari, Bari, Italy
G Mantriota
Dipartimento di Ingegneria e Fisica dell'Ambiente, Università della Basilicata, Potenza, Italy

Chapter 6 – Cylinder Balancing Control of Direct Injection Engines
G N Heslop and *J Dixon*
Visteon UK Limited, (Visteon Automotive Systems), Basildon, UK

Chapter 7 – Continuously Variable Transmission with Electromechanical Power Splitting
G Avery and *P Tenberge*
TU Chemnitz, Germany

Chapter 8 – The Design of a Parallel Hybrid Transmission Control System
J Marco
Pi Technology, Cambridge, UK
R Ball
Warwick Manufacturing Group, University of Warwick, Coventry, UK
R P Jones
School of Engineering, University of Warwick, Coventry, UK

About the Editor

Nick Vaughan BSc(Eng), PhD, FIMechE, CEng, joined the University of Bath in 1978 after working in the aerospace industry at Filton and doctoral research work at the University of Bristol.

His work in both teaching and research at Bath has been principally in control systems, modelling, and simulation. Nick has mainly applied his work to vehicle transmissions, in particular continuously variable transmissions. This work has been with both variable ratio belt and toroidal drives, and in both engine and transmission control. Most recently this has involved a number of projects associated with powertrain integration and control. He has published widely in these and other systems control areas, and is currently co-authoring a book on vehicle transmissions.

Research work has involved continual contact with industry, with direct funded projects from Ford and other international companies, as well as collaborative projects funded by EPSRC, DTI, and the EC. He led the application to obtain EPSRC funding to start an international network in Powertrain Systems and Driveline Technology (TxNet).

Nick has been an active IMechE member for many years. He has been Academic Liaison Officer, is currently an MPDS Mentor, a member of the Degree Accreditation Panel, and a member of the both the Automobile Division Board and Programme Executive.

Foreword

The search for motor vehicles with ever-greater fuel economy combined with greater refinement has encouraged a more holistic approach to many areas of vehicle development. One of the most important of these is the integrated approach to the vehicle powertrain and its study as a complete system from a control perspective. These considerations provided one of the drivers for the formation in 1998 of an EPSRC sponsored Network in Powertrain Systems and Driveline Technology (TxNet). It was this Network that proposed and co-sponsored the event around which these contributions are based, and it continues to act as a focus group for work in this area. This book brings together contributions from around the world all concerned with these integration aspects of powertrain systems.

Jackson of Ricardo Engineers has provided a comprehensive introduction and background to the topic as a whole. This sets the scene in which powertrain engineers have to operate and identifies the main legislative, environmental, and customer demand pressures that are pushing the technology forward. He builds on this base by consideration of emerging technologies to predict future trends. This sets the scene for the remaining papers in this book that can be subdivided into those that deal with control and dynamic aspects, and those that are concerned more with design.

Murray's contribution first overviews the particular characteristics of the Torotrak IVT that are relevant in a control context. He details aspects of both the hardware implementation and the software strategy used in the powertrain control. He also deals with the compromises that are generic to CVT control, principally the conflicts that exist between optimizing steady-state operation and delivering transient response. The other authors also contribute to aspects of this debate. Serrarens describes the development and control of a flywheel system to boost transient system response and gives rise to the concept of a zero inertia powertrain. The vehicle used in this work has been developed as part of the Ecodrive project in The Netherlands and the experimental results demonstrate the benefit. There are two further contributions addressing this aspect that are complementary. Carbone gives details of an analytical approach to determine the limits of performance from belt drive systems, whereas, Kim's work is a proposal for a control solution to get the best from the hardware. In this he develops an integrated engine control algorithm that is optimized to compensate for CVT efficiency and inertia. Heslop provides the final contribution in the control group. He details work at Visteon that is also concerned with engine control having implications for the driveline. This looks at strategies to provide greater balance between the firing of cylinders and hence smoother engine operation. This has a significant contribution to make in terms of improved NVH, particularly in low-speed high-torque regions of operation as identified by Murray.

The other two papers deal with more mechanical integration aspects of powertrains and both are considering hybrid system design. Tenberge's work incorporating two electrical machines in a split path layout provides an elegant solution. This gives the benefits of a CVT in conventional operation but also the extension by providing the features necessary for regenerative braking or drive in a hybrid scheme. Marco describes details of the HERO vehicle developed as part of the UK Foresight Vehicle Programme. This design is principally concerned with characteristics appropriate for a vehicle operating both on and off road. Although the resulting driveline solution is more conventional it nevertheless provides a fascinating solution to a different set of problems.

Finally, my thanks to the authors of the individual sections in this book for their time and effort in preparing their contributions. Also to Kate Lewis of the Institution of Mechanical Engineers for her considerable effort in putting together the Seminar around which this book is based.

Nick Vaughan
BSc(Eng), PhD, CEng, FIMechE
Senior Lecturer – Department of Mechanical Engineering
University of Bath
March 2001

1

Introduction to Advances in Powertrain Technology

N S Jackson

The key market drivers for future powertrains are legislative requirements, fuel economy improvements, consumer demands for performance and driveability and the need for cost control to realize a competitive unit price. Globalization has led to increasing commonality of platforms and powertrains which must be adapted to specific brand identities and niche applications to maintain marque values.

The spark ignition gasoline engine has benefited from many decades of development with the application of significant R&D resources. This has resulted in a refined, highly optimized powerplant with increasing specific performance and exceptionally low exhaust emissions. Conversely, the passenger car diesel engine has received far less attention for a much shorter period. However, while the gasoline engine has remained the dominant passenger car power plant, the diesel engine has benefited from significant improvements in recent years. Research and development resources have been directed towards improvements in refinement, power density and noise characteristics in addition to low fuel consumption. The diesel engine also benefits from not only better thermal efficiency than a conventional port injected gasoline engine but also benefits from improved volumetric fuel consumption due to the c.10–12 per cent increased energy density of diesel fuel.

The key to many of these improvements has been the growth in the capabilities of system controllers and electronic actuation. This will continue in the future with more flexible and powerful control systems coupling the engine, FIE, gas exchange, aftertreatment system, transmission, electric drives, and ancillaries. The introduction of these controllers will not only provide more optimum operation of individual components but enable a system engineering approach to powertrain development.

Improvements in vehicle fuel economy remains one of the critical challenges facing powertrain engineers. Market pressures and impending CO_2 emissions agreements dictate that innovative approaches to the efficiency of combustion engines must continue to be made. Two examples of this approach are the small bore HSDI diesel engine and the Lean Boost gasoline engine.

The HSDI engine features cylinders with reduced swept volume, along with advanced combustion systems, fuel injection equipment, and turbocharging technology. The Lean Boost concept combines direct injection, homogeneous full load lean operation and pressure charging. Lean operation at full load reduces octane requirement and allows a high compression ratio to be used. Boosting is applied to produce an engine output which allows downsizing.

In a vehicle application, the engine is only one part of a more complex system that makes up the powertrain. It has become increasingly important to optimize specific areas of the engine within this context so that for instance, the combustion system can tolerate changes in parameter settings necessary to control exhaust characteristics to maximize aftertreatment system performance. The complexity of the full powertrain system requires sophisticated system simulation tools that can be used to explore interactions between components and subsystems and ensure that the system as a whole operates under optimized conditions rather than as a collection of isolated components.

Although significant progress continues to be made in alternative powertrain concepts such as fuel cells, the overall efficiency and cost-effectiveness of the internal combustion engine will ensure that this will remain the dominant core element of vehicle powertrains for at least the next 20 years.

2

Control of an Integrated IVT Powertrain

S Murray

Abstract

The Torotrak IVT (Infinitely Variable Transmission) uses a torque controlled full toroidal traction drive in a two-regime configuration. Low regime is a split-power system and provides forward and reverse drive. High regime extends the ratio of the transmission in forward drive. The change from low regime to high regime is achieved with the aid of two wet plate clutches that are synchronized during a regime change.

This paper briefly introduces the mechanical system and the characteristics that make the IVT controllable as part of an 'Integrated Powertrain'. Some of the opportunities this affords to the system designer are explored. A brief description of the control hardware and algorithms are presented, along with some discussion of the CAE control system design tools presently employed at Torotrak.

2.1   Introduction

The Series III Torotrak transmission was born from the requirements to design, develop, and demonstrate a new generation of Torotrak IVT, delivering durability and functional attributes in line with customer expectations, along with outstanding fuel economy. To achieve these objectives, in the very limited time-scales, required a change in the control system design methodology used. Power, flexibility, and speed were the keywords.

Advanced floating-point microprocessor hardware with integrated instrumentation, automatically generated code, advanced calibration tools, and integrated test and simulation tools have all been used to ensure success.

2.2   The Torotrak infinitely variable transmission

2.2.1   IVT basics

A sphere between two toroidal surfaces transmits drive in a 1:1 ratio. A slice through the sphere has the same characteristics, but is a highly efficient mechanism for changing ratio.

In the Torotrak Variator an input disc with a toroidal running surface is fixed to the drive shaft. An output disc having matching toroidal surfaces is connected to the final drive. The use of three rollers in the toroidal cavity and the introduction of a second cavity means that there are equal and balanced forces within the variator without any efficiency loss. The angle of the roller equates to the input/output speed ratio.

Hydraulics are used to vary contact forces within the whole assembly, enabling rapid torque changes to occur. In the variator the discs and rollers are cooled and lubricated with traction fluid, which provides the drive medium.

The system layout and power flows are shown in Fig. 2.1. Power is transmitted to the driven wheels through a mixing epicyclic gearset. The transmission input is split into two paths. One drives the input discs of the variator. Via the roller this drives the output discs, which in turn drives the sun of the epicyclic gearset. In low regime the other drives the planet carrier of the epicyclic (through the low regime clutch). The annulus is connected by a shaft to the driven wheels. The epicyclic gearset is bypassed (via the high regime clutch) as the transmission shifts into high regime.

This combination of epicyclic and variator does away with the need for an inefficient starting clutch or torque converter. As the variator ratio changes in low regime the transmission provides ratios from reverse through 'geared neutral' to synchronous. Geared neutral is unique to the split-power design, and occurs when the vehicle wheels are stationary, and the engine is rotating, while a direct mechanical link exists between them. At synchronous ratio the relative speed across both high and low regime clutches is zero, allowing a seamless transition into high regime where the variator ratio is swept once more to transition from synchronous ratio to high overdrive.

The wide ratio spread, high overdrive capability, and rapid transient response maximizes the control systems ability to optimize the efficiency of the entire powertrain. Hence, the delivery of driveability with greatly improved fuel economy and emission compliance.

2.2.2   The Series III IVT

Present efforts are focusing on the Series III Torotrak IVT, with which it is intended to demonstrate the maturity of the technology. It is targeted at the US Sport Utility Vehicle (SUV), which is a high-volume market (around 7 million per annum).

Few alternative technologies exist to improve fuel economy in this segment, due to the large capacity, high-torque engines, which are presently outside the capabilities of alternative CVT technologies.

Series III mechanical design was completed in conjunction with Ricardo Midland Technical Centre by Q2 99. and is presently undergoing control system development and durability prove out.

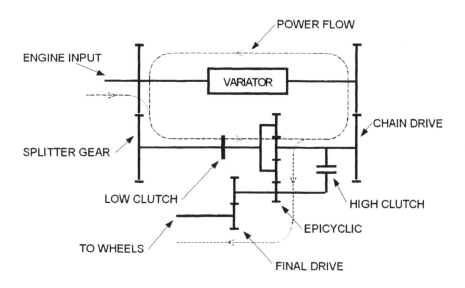

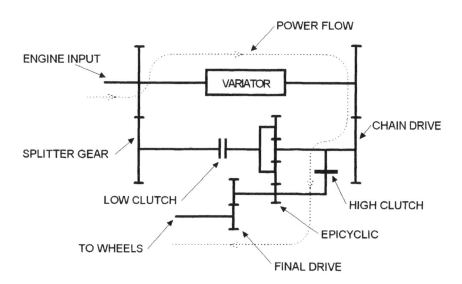

High Regime Drive Power Flow

Fig. 2.1 IVT layout and power flows in each regime

2.3   Control of the Torotrak IVT

2.3.1   Integrated powertrain control

The Torotrak IVT is not a direct replacement for a conventional transmission of either stepped or continuous ratio type. Its unique characteristics demand that the powertrain is controlled as a single system, rather than as separate engine and driveline components.

2.3.1.1   Torque control

A simplified one-roller model in high regime is shown in Fig. 2.2(a), with the engine and vehicle inertias acting upon the variator. The Torotrak IVT actually contains six rollers; two banks of three in axially balanced cavities. The associated speeds of the two inertias will determine the speed of the roller. Applying a force to the roller will accelerate or decelerate the two inertias, depending upon the direction of the reaction force. This will change the speed of the engine and/or vehicle resulting in a change of variator ratio. The application of a castor angle to the roller axis provides a self-steering force that allows it to find its own ratio. The flow process loop for torque control is shown in Fig. 2.2(b).

The roller force balance for one roller is shown in Fig. 2.2(c). If hydraulic pressure is applied to the reaction piston, the force is reacted tangentially resulting in a torque on each disc. The rollers can only transmit torque when a reaction force is applied. If the reaction force is zero then the rollers merely roll between the discs. Thus it is the torque which is the control variable rather than the speed ratio as in other CVTs.

Given this torque control effect it is apparent that the control medium in the hydraulic system is pressure rather than flow, as may be expected initially. This is achieved by controlling the pressure on each side of an actuator, an imbalance of which will cause a reaction force on the variator.

2.3.1.2   Powertrain torque balance

To set a torque at the drive wheels, a reaction torque is required on the variator. While this changes with ratio and regime, as a function of the transmission geometry, it is constant in steady-state conditions. This reaction torque, proportional to actuator pressure, also generates a proportional transmission input torque, or engine loading. If steady-state conditions are to be maintained then the torque that the engine is generating and the torque that the transmission is loading must be equal. Any imbalance will cause the engine speed to change.

The torque-controlled nature of the full toroidal variator has effectively de-coupled the engine from the wheels. Rather than the wheel torque being controlled by the engine output, or throttle position, it is instead controlled by variator reaction torque.

2.3.2   Control hardware architecture

The Series III control hardware architecture was designed to allow the simplest possible conversion between the various SUV platforms. Presently, conversions have been made on the Ford Expedition and the GM Silverado.

A simplified representation of the control hardware configuration is shown in Fig. 2.3. Three electronic control units are used. The supervisory Powertrain Control Unit (PCU) controls the engine management system and electronic throttle.

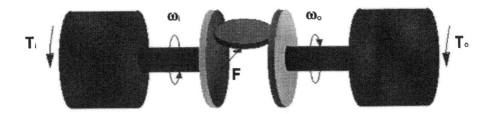

(a) Simplified one roller variator model

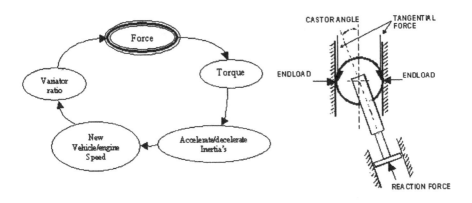

(b) Torque control causality process flow   (c) Single roller force balance

Fig. 2.2 Torque control of the IVT

2.3.3 Engine controller

Persuading vehicle manufacturers to modify production engine control strategy has been difficult for Torotrak in the past, inevitably leading to confidentiality issues and prolonged development periods.

Replacing the conventional engine control unit is a radical way of removing these issues, as it means a new base engine calibration for each target vehicle. To offset this, significant optimization of the original module's calibration would be required to match the requirements of the IVT, so the incremental calibration work is minimized. The use of a common engine controller across all vehicle platforms means engine controller strategy modifications are required only once and the engine calibration tool and techniques remain consistent.

The OEM's controller is replaced with an EMS 5.3 engine management unit from Zytek Systems Limited. This unit utilizes Zytek proprietary strategy for fuelling, ignition, and canister purge, with a modular interface provided on the Controller Area Network (CAN) bus to allow PCU intervention, as required.

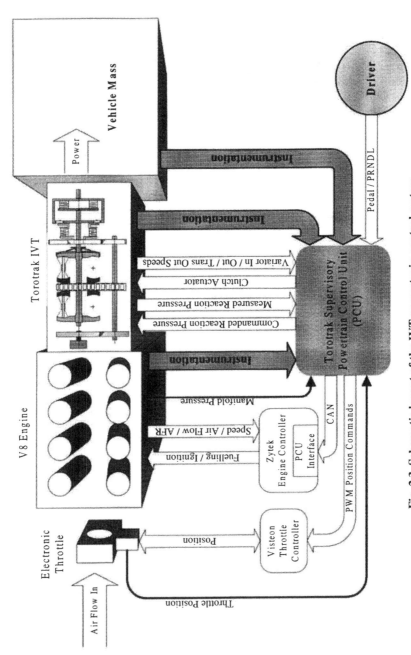

Fig. 2.3 Schematic layout of the IVT powertrain control system

2.3.4 Electronic throttle

Electronic throttle control is required to enable the link between engine output torque and driver's pedal to be removed. Visteon produces an electronic throttle with an integrated controller, which is now in production on the Jaguar S type. Again, this unit has been adopted across all vehicle platforms.

2.3.5 Powertrain control module

On the earlier Series II transmission development programmes Torotrak utilized a modified Lucas GEMS engine management controller, based on the Intel C196 processor and the C programming language. In order to support the Series III programme, with its aggressive time-scales and targets, a new control system development environment was specified.

Initial development proceeded using the dSPACE rapid prototyping environment, and custom-generated interface electronics. This was, however, an interim solution, providing rapid prototyping benefits but at high cost and with limited functionality and reliability. An updated specification was drafted:

- Powerful floating-point main processor
- Second microprocessor for main processor monitor/limited operation strategy
- Robust enough for vehicle durability and fleet trial (inc. EMC)
- Automotive temperature specification (-40°C to +80°C)
- Extensive and flexible integrated i/o to allow both instrumentation and powertrain control interfacing
- Integrated three axis accelerometers and temperature monitoring
- Built in data logging to support fleet trial and development
- High-performance calibration interface

'Proteus', an advanced automotive development ECU, designed and manufactured by Prodrive, met the key features of Torotrak's specification. It is based on two MPC555 floating-point micro controllers that Motorola designed specifically for automotive applications. Prodrive have added extensive flexible i/o and memory.

Internally the 'master' processor, communicating with the 'slave' unit to access its i/o, manages the control function. Safety critical inputs are shared by both processors via independent i/o, allowing the 'slave' monitor capability.

Calibration is effected using the Vision Hub and software from Accurate Technologies Inc, providing a high bandwidth interface over a dedicated CAN bus, running CAN Calibration Protocol (CCP).

2.4 Controller software development

The following section considers how the supervisory PCU software is implemented on the Proteus controller. A simplified closed loop controller has been used as an example of steps involved in generating target software, and how this is implemented and calibrated in vehicles. The actual PCU implementation is then briefly reviewed.

2.4.1 Example controller implementation

The reaction torque on the variator determines the wheel and engine torque, as described previously, and this in turn is achieved using closed loop pressure control. Two flow-control valves independently raise pressure on each side of the roller pistons, applying force in either direction.

A simplified version of the controller block diagram used to achieve this functionality is shown in Fig. 4.4. At the core of the controller is the discrete PID block. Below its mask is a resetable discrete controller algorithm, written in C. The block is stored in a Torotrak software library, and is reused in several parts of the PCU.

Calibration parameters are shown as global variables. For example KP_SERVO, TI_SERVO, and TD_SERVO which are the PID controllers gain and time constant inputs. The default values of these parameters are inherited at code build time from the Matlab workspace, where they have been placed by running a calibration script ('.m') file. Look-up tables, and functions are also set up this way. As this calibration file is a Matlab script, much more powerful features may be exploited, such as discrete filters represented in the example by the discrete transfer function block containing $\dfrac{PRESS\_SERVO\_FILT\_NUM}{PRESS\_SERVO\_FILT\_DEN}$. An extract of the '.m' file calibration script used to generate some of these constants is shown in Table **2.1**

Table 2.1 Example extract from calibration script

```
%%%%%%%%%%%%%%%%%%%%%%%%%%%%%%%%%%%%%%%%%%%%%%%
%
% SERIES III CALIBRATION FILE
%
% Vehicle number 29/11 last revision 20/4/00
%
%%%%%%%%%%%%%%%%%%%%%%%%%%%%%%%%%%%%%%%%%%%%%%%

LOOP_TIME = 0.002;     % Sample time

KP_SERVO = 0.75;       %Pressure controller gain (bar per bar)
TI_SERVO = 0.05;       %Pressure controller integrator time constant (seconds)
TD_SERVO = 0.01;       %Pressure controller derivative time constant (seconds)

% Pressure filter design algorithm, 4th order butterworth Wn=40Hz

[PRESS_SERVO_FILT_NUM , PRESS_SERVO_FILT_DEN] = butter(4,40/(1/(2*LOOP_TIME)));

% Variator reaction torque to reaction pressure look up table

TRQ_REAC_VAR2PRESS_DIFF_X = [0 1357]; %Variator reaction torque envelope (Nm)
TRQ_REAC_VAR2PRESS_DIFF_Y = [0 47.9]; %Pressure differential envelope (Bar)
```

2.4.1.1 *Proteus I/O interface*

The transition from the prototype implementation on dSPACE to the Proteus environment was made relatively simple. Prodrive have provided a replacement for the dSPACE RTI (Real Time Interface) blockset to give access to the Proteus i/o.

Proteus analogue input and PWM output blocks are shown at the left and right hand side of Fig. 4.4. While these are normally contained in separate input and output conditioning blocks in the PCU, they are shown in this block by example. Analogue inputs are converted to engineering units before being passed to the control strategy.

Similar blocks are available to handle digital and pulse inputs, and digital, analogue, PWM and current controlled outputs. The availability of i/o in the Simulink environment makes the connection and set up of new sensors and actuators straightforward.

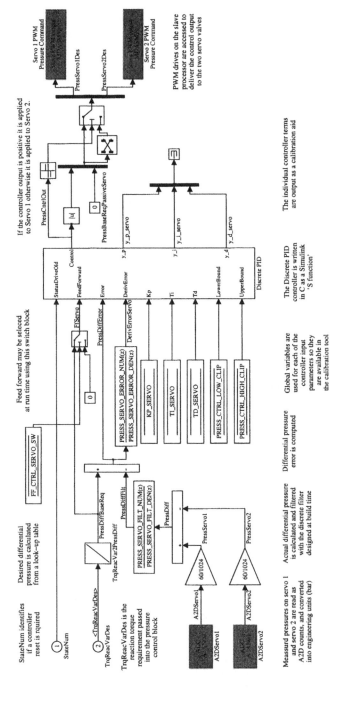

Fig. 2.4 Simplified pressure controller implementation

2.4.1.2 *Generating and running the object file*

The model is converted to C code using the Mathworks Real Time Workshop Build command from the Simulink pull down menu. RTW invokes the 'Target Language Compiler' which translates the model into compileable source code. A specific 'template make' file then controls the generation of the '.elf' object file.

An in-house 'RobGen' utility, extracts label and symbol information from '.elf' and '.rtw' files, and generates a data dictionary containing variable and calibration address information. This is automatically filled with default ranges and engineering units based on a standard variable naming convention. A documentation database is used to add notes and additional parameter information to the '.rob' file generated, which can be used on line in the calibration environment.

The '.rob' and '.elf' files are loaded into the ATI Vision hub via a USB connection to the PC. CAN Calibration Protocol (CCP) resident in the BIOS of the Proteus controller is used to down load the object file from the ATI unit over the dedicated CAN link, and programme the flash memory.

2.4.1.3 *Calibration*

Flash programming is only required once, and when this operation is completed the control strategy will run with or without the connection of the ATI tool, using the default calibration loaded at build time. When connected, the Vision unit provides a powerful calibration environment with maximum data throughput and logging of over 20 000 variables per second, limited only by disk space. The '.rob' file is used as the controller parameter map, and allows access to all variables, calibration parameters, tables, and maps identified by name in the controller diagram.

2.4.2 Control strategy architecture

The techniques demonstrated in the example above are used exclusively to generate the 'master' processors supervisory control software from a high level Simulink and Stateflow block diagram representation of the control requirements. A representation the controller's primary functionality is shown in Fig. 2.5. The control flow from left to right is described across the bottom of the figure.

There follows a brief description of the implementation of the functions that interpret driver, fuel economy, and emission requirements. Other areas of particular interest such as dynamic powertrain matching and engine/transmission modelling in the controller are beyond the scope of this paper.

2.4.2.1 *Interpreting driver requirements*

The driver has two main inputs to the controller – the drive select lever and the accelerator pedal. All of the driver's primary requirements for vehicle speed and acceleration, and engine speed and acceleration are interpreted from these. While attempting to generate the driver's requirement, the controller must also: minimize fuel consumption; comply with emission requirements; and avoid operating areas of excessive noise, vibration, and harshness (NVH).

The driver's needs are considered separately during steady-state and transient operation. Steady-state conditions are deemed to be occurring based primarily on accelerator pedal speed. That is to say, beyond a certain rate of change of pedal position a 'transient event' is occurring, otherwise the system is in steady-state driver requirements mode. Note that this

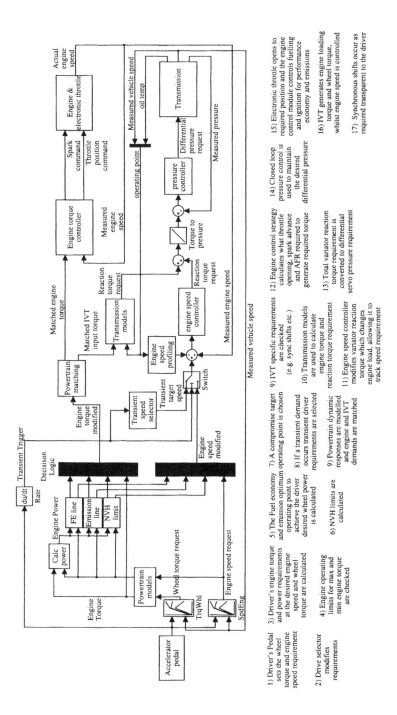

Fig. 2.5 Torotrak IVT powertrain control schematic

1) Driver's Pedal sets the wheel torque and engine speed requirement

2) Drive selector modifies requirements

3) Driver's engine torque and power requirements at the desired engine speed and wheel torque are calculated

4) Engine operating limits for max and min engine torque are checked

5) The Fuel economy and emission optimum operating point to achieve the driver desired wheel power is calculated

6) NVH limits are calculated

7) A compromise target operating point is chosen

8) If a transient demand occurs transient driver requirements are selected

9) Powertrain dynamic responses are modelled and engine and IVT demands are matched

9) IVT specific requirements are checked (e.g. sync shifts etc.)

10) Transmission models are used to calculate engine torque and reaction torque requirement

11) Engine speed controller modifies variator reaction torque which changes engine load, allowing it to track speed requirement

12) Engine control strategy calculates what throttle opening, spark advance and AFR required to generate required torque

13) Total variator reaction torque requirement is converted to differential servo pressure requirement

14) Closed loop pressure control is used to maintain the desired differential pressure

15) Electronic throttle opens to required position and the engine control module controls fuelling and ignition for performance economy and emissions

16) IVT generates engine loading torque and wheel torque, whilst engine speed is controlled

17) Synchronous shifts occur as required transparent to the driver

does not necessarily mean that the driver requires constant speed. For example, a constant full pedal launch may be classified as a steady-state condition, even though the vehicle condition is rapidly changing.

2.4.2.1.1   Steady-state drivers requirements
The driver's requirement is split into a target wheel torque and engine speed. This allows the launch and cruise characteristics to be matched to a standard vehicle, and ultimately to improve areas of poor driveability. The vehicles performance is perceived to be 'correct' if the driving force or wheel torque and the engine speed are correct, or more accurately, if the time history of these parameters is as expected.

A simplified representation of a wheel torque versus vehicle speed for a family of typical launches from rest of a torque converter automatic is shown in Fig. 2.6(a). The shift events that normally apply steps to this characteristic have been smoothed out. In an attempt to emulate this characteristic the current IVT implementation has four look-up tables for each drive selector lever position, a pair for wheel torque, and another pair for engine speed. The total steady-state driver requirement, Fig. 2.6(d), is the sum of the 'launch' and 'cruise' tables as shown in Fig. 2.6(b) and 2.6(c). The IVT and standard auto characteristics (Fig. 2.6(a) and 2.6(d)) are comparable with the exception of the first few kph of vehicle speed, which is taken care of by the transient strategy.

2.4.2.1.2   Transient driver requirement
The entry and exit of the transient state in normal operation is decided exclusively by the pedal speed (rate of change of pedal position). When there is a significant change in pedal position over a short period of time, a transient is occurring, and the target engine speed and wheel torque calculation method is revised.

In steady-state conditions the assumption is made that the rate of change of engine speed is not significant, i.e. the transmission input torque and engine output torque are balanced. During a transient event the engine is required to follow a profiled acceleration, which emulates conventional transmission ratio shift. In order to achieve this, energy is required to accelerate the engine and transmission input components. This requirement can be satisfied by changing either the engine output or transmission input torque, or both in combination.

The interpretation of the driver's requirement during a transient event, then, does not directly include a wheel or engine torque requirement. Instead, it involves controlling the rate of change of engine speed while targeting a new steady-state operating point, which may also be dynamically changing. The wheel torque is then a consequence of the acceleration applied to the engine, and the throttle set point to achieve the new target operating power. In this way the transient behaviour can be configured between gentle shifts and rapid 'kick down'.

Small engine speed errors would build up when engine output torque and transmission input torque estimates are incorrect. The torque imbalance is effectively integrated by the inertias of the engine and transmission input components. These are avoided by using closed-loop engine speed control, and models of the system inertias across the operating envelope.

2.4.2.1.3   Effect of transient driver requirement on launch characteristic
Figure 2.6(e) shows what happens to the steady-state characteristic of the IVT's output torque tables after the transient strategy has manipulated it during a tip-in from rest. As soon as the transient is triggered the engine speed profiling begins, and the engine speed is controlled to

track toward a new steady-state requirement. This, in turn, requires torque to be used to accelerate the engine toward the target speed, which reduces the torque available to transmit to the wheels. When the target speed is achieved the system reverts back to steady-state control. Overall, the effect produced is similar to that of the conventional torque converter automatic.

2.4.2.2 Fuel economy and emission control lines
A two-dimensional table exists for each of the controlled parameters, HC, CO, NO_x, and fuel consumption, which describes the 'control line'. This is a line of optimum performance through the engine speed torque map for the specified component. Figure 2.6(f) shows an example for NO_x.

With an integrated powertrain control philosophy, other parameters such as closed-loop engine controller bias, spark advance, EGR flow, etc., can be used to manipulate engine fuel consumption and emissions. It is intended that these secondary input parameters will be explored over the coming months.

Each table generates an optimum torque/speed at which to produce the present engine power requirement, from the driver's requirements tables described earlier. Weighting functions are applied to each of these and output of the driver's requirements, based on the present accelerator pedal position. The final speed/torque set point is calculated from the weighted average of the requirements, using the driver's engine power requirement from the powertrain models. This ensures that whatever engine speed and load is selected after the control lines have been used, the wheel torque anticipated is still generated.

2.4.2.3 NVH control
The final manipulation of the engine speed and load set point is in place to avoid areas of poor NVH. Generally, NVH occurs at low speed and high load, and can introduce a severe restriction on the ability of the system to deliver economy. The control strategy chosen here has final authority over the minimum speed and maximum load at which the engine is allowed to operate in steady-state conditions. However, excursions into areas of poor NVH are allowed, dependent on a weighting factor based on the accelerator pedal. By way of example, full throttle at 1000 rpm might not be an acceptable steady-state operating condition, however it would be during a launch, where it would be passed through quickly. This strategy allows maximum performance to be extracted without compromising overall driveability.

2.4.3 Slave processor architecture
The 'slave' processor is used to monitor the actions of the 'master' ensuring that the rules set out in the primary requirements specification are not broken. The present implementation is prototyped in Simulink also, though the intention is to implement this in high-integrity MISRA (Motor Industry Software Reliability Association) compliant C code, enhancing the robustness of the entire system to software or controller failure.

In future iterations it is intended to implement elements of core functionality within this high-integrity code, which will be given ultimate authority over the controller outputs. This will provide limited operation capability in the event of critical failures of the 'master' or its i/o or memory.

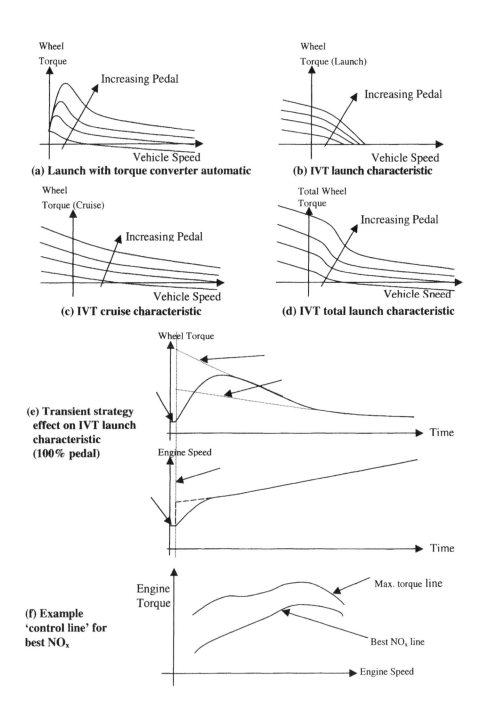

(a) Launch with torque converter automatic

(b) IVT launch characteristic

(c) IVT cruise characteristic

(d) IVT total launch characteristic

(e) Transient strategy effect on IVT launch characteristic (100% pedal)

(f) Example 'control line' for best NO$_x$

Fig. 2.6 How the IVT satisfies the driver's requirements

2.4.4 Full vehicle simulation

Torotrak has a full suite of powertrain modelling blocks also created in Simulink. Particular attention has been given to the precise modelling of the core IVT technology, but the models also comprise full and comprehensive control hydraulics, engine, and vehicle representations. To the base Series III model, neural networks have been added, and used to model engine emissions and fuel consumption.

As the controller and vehicle models are both built in the same simulation language it has been a relatively simple process to produce a 'bench-test' model. Proteus i/o is replaced with links to the models. A simulated 'robot' driver with an accelerator and brake pedal can be used, or data acquired from vehicle testing can be run in the same environment to diagnose software or strategy problems.

As yet the full vehicle simulation is not completely validated, so any results obtained still require vehicle test for confirmation. However, this is already proving to be a valuable tool for testing the effects of strategy and hardware modifications on fuel consumption and feedgas emissions, without the need for expensive vehicle tests.

2.5 Acknowledgements

The author would like to thank Tarragon Embedded Technology Limited for their continued support in the development of Torotrak's control systems. Particular thanks go to Geoff Frost who has been instrumental in the present implementation from the outset of the project.

2.6 References

James, I. and **Price, D.,** (1999), Modelling Techniques Applied to the Development of Torotrak's Series 3 IVT, *CVT99,* Eindhoven.

Hough, M., (2000), The Torotrak IVT – Forward to Production, *Global Powertrain Congress, GPC2000,* Detroit.

3

Driveability Control of the *ZI*® Powertrain

A F A Serrarens

Abstract

Vehicle powertrains with a continuously variable transmission (CVT) have a large freedom in controlling the engine speed and the torque at the wheels. Due to rotating inertias within the engine and transmission, the response of the vehicle during large engine speed shifts may appear reluctant or even counteractive. To overcome this behaviour, the CVT is augmented with a planetary gear set and a compact steel flywheel. The new transmission seamlessly combines two contradictive features: the driveability in terms of pedal-to-wheel response is greatly improved, and a large leap towards optimal fuel economy can be made. The latter is achieved by cruising the vehicle at extremely low engine speeds as of the large ratio-coverage of the CVT. The flywheel acts as a 'peak sheaver': it delivers power during engine acceleration and absorbs kinetic energy during engine decelerations. In this chapter, design and control aspects of the new powertrain are discussed. Results from simulations and preliminary experiments show the improvements, both in driveability and fuel efficiency.

3.1 Introduction

The combination of an internal combustion engine and a continuously variable transmission (CVT) in vehicular powertrains should be designed such that, the desired vehicle response can be achieved at the highest possible efficiency for all vehicle loads and speeds. CVT's offer a large freedom in controlling the engine speed independently from the vehicle speed. Here, a Van Doorne's metal V-pushbelt CVT is considered. The ratio coverage of this CVT is considerably larger than that of stepped automatic or manual transmissions. This paves the way for a high-fuel economy of the powertrain. The highest ratio of the CVT, better known as the *OverDrive* ratio, is able to push the engine speed to extremely low values. To maintain the current speed of the vehicle, the engine torque generally has to be quite high. This brings the engine operating points in a more fuel economic region, see (**7**).

In Fig. 3.1, the operating points of a 1.6 ltr internal combustion engine with a 4-AT and a CVT are compared for a fragment of the FTP drive cycle. The figure clearly illustrates the influence of the high OverDrive ratio in a CVT powertrain, forcing the operating points to the upper left region of the engine map. Ideally, all operating points should lie on the so called E-line, i.e. the collection of operating points with the lowest possible brake specific fuel consumption. Figure 3.1 also illustrates that, even with the large OverDrive ratio of the CVT, it is not always possible to track the E-line. However, the additional fuel savings that can be gained by further enlarging this ratio are quite small. The currently available OverDrive ratio realizes fuel savings up to 15 per cent with respect to the 4-AT measured on a NEDC drive cycle.

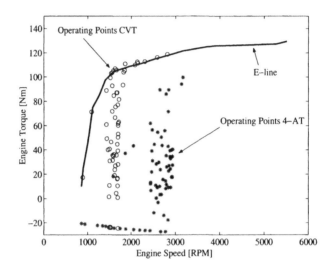

Fig. 3.1 Engine operating points with 4-AT and CVT

3.1.1   The driveability versus fuel economy problem

If the OverDrive ratio is used to track the E-line at stationary vehicle speeds (cruising), the driveability deteriorates considerably. This is typical for *any* transmission with a large ratio coverage. If a large acceleration of the vehicle is requested by the driver, a large leap in engine speed has to be made by shifting the transmission to a low ratio. The powertrain inertias then burden the instant increase of wheel power. To illustrate this further, Fig. 3.2 shows an engine map with four operating points at the engine power hyperbolas P_e, P_e', P_e'', and P_e^{max}.

The engine torque is bounded by the Wide Open Throttle line WOT whereas the engine speed is bounded by the dashed line. For operating points on the steep part of the E-line (e.g. point 1) a moderate step (towards point 2, say) in the desired output power can be realized by an increase of the engine torque T_e. For operating points on the flat part of the E-line, however, the potential increase of T_e is small and a higher output power can only be delivered if the engine speed is increased. In the figure this is illustrated by a transition from point 2 towards point 3 or even to point 4, where maximal combustion power is realized. The inertias of the engine and the transmission will seriously counteract the net output power of the powertrain during such speed transients. More specifically, the resulting wheel power is restrained for relatively slow CVT shifting and even becomes negative for fast shifting.

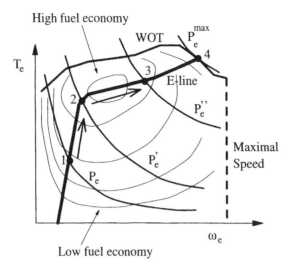

Fig. 3.2 Engine map

In practice these responses are improved by choosing a lower CVT ratio than OverDrive during cruising. The consequences are a higher engine speed and a lower engine torque (see Fig. 3.1). In the case of a kick-down the required engine speed excursion will thus be smaller and the step towards the WOT torque is larger, leaving more instant power for accelerating the engine and the vehicle.

3.2   Mechanical solution: zero inertia powertrain

Although the mentioned driveability problem can be solved relatively well in practice, it cannot, in general, be combined with a high-fuel economy because only engine speeds as low as possible will – at least during cruising – result in a high engine efficiency. It is desirable that the OverDrive ratio is maintained during cruising and that the engine obtains a short-term torque assist during a fast speed transient. Solutions to this problem generally use an electric motor assist (**9**) and (**10**). At least 40 to 50 kW is shortly required to solve the driveability problem. Although electric motors can be overloaded by a factor of two to three, a large battery pack or ultra-capacitor is required to supply this power. This makes the system relatively expensive, voluminous, and heavy. Here, a purely mechanical solution without these drawbacks is chosen. In this solution, a compact steel flywheel is connected in parallel to the CVT via a planetary gear set. The idea is to cancel the primary powertrain inertias by another inertia. Therefore, the new powertrain is coded Zero Inertia (ZI) powertrain. A chosen connection of the planetary gear set to the CVT makes it possible to have opposite signs of engine and flywheel accelerations. Consequently, the flywheel boosts the powertrain during CVT downshifts and absorbs kinetic energy of the powertrain during upshifts.

3.2.1   Functional design of the powertrain
The connection of the planetary gearset and the flywheel to the CVT can be realized in many variants:

1. CVT between engine and wheels, flywheel in parallel;
2. CVT between engine and flywheel, wheels in parallel;
3. CVT between wheels and flywheel, engine in parallel;
4. planet carrier connected either to wheels, flywheel, or engine;
5. sun gear connected either to flywheel, engine, or wheels;
6. annulus connected either to engine, wheels, or flywheel.

A first optimization learned that variant 1 was the best option since:

*   it is closest related to a conventional CVT powertrain, so the available extensive knowledge can be adopted;
*   blind powers are zero for stationary situations and relatively low during transients. This is not true for variants 2 and 3;
*   contrary to variant 2, the required flywheel inertia is low;
*   contrary to variant 3, the accuracy of the CVT ratio control does not have to be higher than for a conventional CVT powertrain.

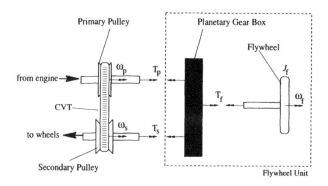

Fig. 3.3 Functional layout of the ZI powertrain

3.2.2 Principal of operation

The layout of the ZI powertrain is given in Fig. 3.3. The gearbox and the flywheel (speed ω_f, moment of inertia J_f) together form the *flywheel unit*. The CVT (with gear ratio, i_{CVT}) relates the primary speed, ω_p, and secondary speed, ω_s by

$$\omega_p = \frac{\omega_s}{i_{CVT}}, \tag{3.1}$$

where i_{CVT} is bounded by a lower value, i_{low}, and an OverDrive value, i_{OD}, so

$$0 < i_{low} \leq i_{CVT} \leq i_{OD}.$$

For a functional description of the powertrain, it is irrelevant which of the variants 4, 5, and 6 is chosen. This choice has to be made in the mechanical design optimization. In the functional design, the planetary gearset is considered as a three-pole gear box which relates ω_p, ω_s, and ω_f by:

$$\omega_f = i_s \omega_s + i_p \omega_p \tag{3.2}$$

The gear ratios i_p and i_s are constant. Using equation (3.1) this relation for ω_f can be rewritten as

$$\omega_f = \left(i_s + \frac{i_p}{i_{CVT}} \right) \omega_s . \tag{3.3}$$

Hence, the flywheel speed, ω_f, depends on the CVT ratio, i_{CVT}, and the secondary speed, ω_s, (or, taking the final reduction into account, the wheel speed). There is one unique CVT ratio that results in $\omega_f = 0$ for $\omega_s \neq 0$. This so-called *geared neutral* ratio, i_{GN}, is given by $i_{GN} = -i_p / i_s$.

The relevant torque within the flywheel unit satisfy

$$T_p = i_p T_f ; \tag{3.4}$$
$$T_s = i_s T_f ; \tag{3.5}$$
$$T_f = -J_f \dot{\omega}_f . \tag{3.6}$$

The equations (3.4) and (3.5) show the torque-split property of the planetary gearset. Using i_{GN} and the abbreviation $J_f^* = i_s J_f$, the relations (3.4) and (3.5) can be rewritten as

$$T_p = i_{GN} J_f^* \dot{\omega}_f \tag{3.7}$$
$$T_s = -J_f^* \dot{\omega}_f , \tag{3.8}$$

The torque, T_{net}, *exerted by the flywheel unit at the secondary pulley* becomes

$$T_{net} = T_s + \frac{T_p}{i_{CVT}} = -\left(1 - \frac{i_{GN}}{i_{CVT}} \right) J_f^* \dot{\omega}_f . \tag{3.9}$$

Assuming that i_{GN} is smaller than i_{OD} the functionality of the flywheel unit can be formulated as:

If the flywheel is decelerated then the torque T_{net} boosts the secondary pulley – and therefore the wheels – as long as i_{CVT} is larger than i_{GN}.

After this formulation of the functionality in terms of the boost torque, T_{net}, the functional design can be optimized for the complete operating range of the vehicle and the engine. Therefore, optimal values for i_{GN} and J_f^* have to be determined.

The flywheel unit is introduced to solve the driveability problem. More specifically, the unit has to deliver extra power to the powertrain whenever large, fast changes in the engine speed have to be made in order to realize significant changes in the wheel power. For this purpose, it must be required that the gear ratio, i_p, in equation (3.2) is negative. Then for $\dot{\omega}_s = 0$ accelerations of the engine (down-shifts of the CVT) are accompanied by decelerations of the flywheel and kinetic energy is transferred from the flywheel to the primary inertia, J_p (all inertias of the engine up to and including the primary pulley). Furthermore, the flywheel will absorb kinetic energy from the primary inertia whenever it is accelerated, that is during CVT up-shifts.

The torque assist mechanism of the ZI powertrain is slightly different from those where the engine is directly assisted, e.g. by an electric motor on the crankshaft. The total torque, T_{tot}, at the secondary pulley is the sum of the net engine torque and the net torque from the flywheel unit, so

$$T_{tot} = \frac{T_e - J_p \dot{\omega}_p}{i_{CVT}} + T_{net} \tag{3.10}$$

where T_e is the combustion torque. It is assumed here that the launching device (*i.c.* the torque converter) is locked. Roughly stated, the combustion torque is used to accelerate the primary inertia whereas the assist torque, T_{net}, is used to accelerate the vehicle. Preferably, i_{GN} is chosen below i_{low} because then T_{net} will assist the powertrain for all possible CVT ratios. In practice, the CVT ratio is much larger than i_{low} except during launch. Choosing i_{GN} somewhat above i_{low} is favorable because then the inertia and thus the size and weight of the flywheel can be chosen smaller. For different values of the design parameters i_{GN} and J_f^* numerous kick-down simulations, well scattered within the operating range of the vehicle and engine, were performed to investigate the initial vehicle acceleration response. The results revealed an optimal choice for these parameters in the sense that J_f^* is as small as possible and i_{GN} is close to i_{low}.

3.3   Control

The ZI powertrain requires a control system for the combustion torque and the CVT ratio to minimal fuel consumption and maximal driveability. For the ZI powertrain only the latter is of importance in real life traffic situations. The combustion torque is controlled using an electronic air throttle whereas the CVT ratio is controlled by manipulating the hydraulic pressures on the pulleys.

3.3.1   Control hierarchy
The control hierarchy (see Fig. 3.4) consists of three layers. The first layer is the driver manipulating the drive pedal to control the vehicle's position, speed, and acceleration with respect to other vehicles. The second layer translates the drive pedal motions into setpoints for the combustion torque and the CVT ratio. The philosophy behind this controller is that the driver is well able to control the vehicle's position, velocity, and acceleration if the error between the desired wheel power (related to the drive pedal position) and the realized wheel power is small for all times. This has to be realized under the constraint of minimal fuel

consumption (E-line tracking) during (quasi-) stationary situations. The third layer controls the engine throttle position and CVT ratio such that the setpoints (given by the second layer) are realized as good as possible. Results of the CVT ratio control strategy can be found in (**8**).

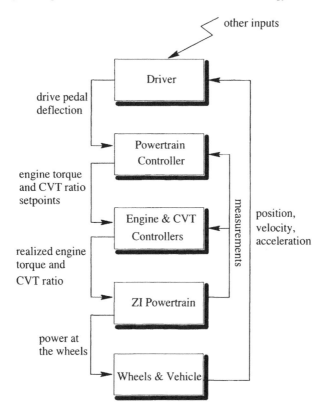

Fig. 3.4 Hierarchy of systems and controls

The remaining blocks in Fig. 3.4 refer to the actual powertrain hardware, wheels and vehicle.

3.3.2   Simulation model
To validate the controller, a simulation model is built and experiments with a realized ZI powertrain are carried out. The simulation model is a seven degrees-of-freedom system describing:

- second-order engine dynamics;
- first-order CVT shift dynamics;
- powertrain inertias and kinematics;
- driveshaft flexibility;
- wheel slip;
- rolling resistance and air drag.

3.3.3   Controller design

In this section the powertrain control strategy is outlined. Given the drive pedal position, this strategy determines the optimal values for the CVT ratio and combustion torque at every time instant such that:

- longitudinal driveability is maximal during engine power transients towards a desired stationary power;
- fuel economy is maximal in (quasi-) stationary situations.

A simple model is used for control design. This model can effectively and unambiguously account for the control objectives.

3.3.4   Driveability control with a fuel economy constraint

The task of the powertrain controller is to find the optimal courses in time for T_e and ω_p such that:

- $$\left\| 1 - \frac{P(t)}{P_d(t)} \right\|_n \leq \xi_P \ \ for \ all \ t \ \wedge \ \ P(t) = P_d(t) \ \ for \ t \to \infty \tag{3.11}$$

- $$\left\| 1 - \frac{T_e(t)}{T_{E-line}(P(t))} \right\|_n \leq \xi_T \ \wedge \ \left\| 1 - \frac{\omega_e(t)}{\omega_{E-line}(P(t))} \right\|_n \leq \xi_\omega \ \ if \ \left\| 1 - \frac{P(t)}{P_d(t)} \right\|_n \leq \varepsilon \tag{3.12}$$

where $|\bullet|_n$ is some norm and ξ_P, ξ_T, ξ_ω, and ε (with $\varepsilon \leq \xi_P$) are small positive numbers. The first requirement loosely formulates the maximal driveability objective whereas the second requirement refers to the maximal fuel economy constraint. Driveability improves for smaller values of ξ_P. The penalty on meeting the fuel economy constraint (tracking the E-line) is high if ξ_T and ξ_ω are small and ε is large. For such, ε, the fuel economy constraint should be met long before the desired power, P_d, is realized.

Translating the above criteria into control actions is deemed to be very difficult. Instead, a relatively simple controller is derived based on a simple nonlinear first order model of the controllable drive power and the idea to separate the generation of setpoints for T_e and ω_p for either driveability or fuel economy. Up to now, this approach is found nowhere in literature, (6). Here, a controller is derived where ω_p, or possibly its derivative, determines the driveability and T_e the fuel economy. Fairly straightforward, it is also possible to design a controller in which the role of T_e and ω_p is interchanged. Even alternating the mutual role of these two quantities is conceivable.

3.4   Results

In this section results from simulations with an advanced validation model and experiments with a test rig are discussed. The performance of the controllers is validated, excluding the influence of the driver. Experiments, including the driver, will be performed in the near future after the implementation the ZI powertrain in a commercially available car. For the moment, an arbitrary though sufficiently exciting pattern in time is chosen for the deflection of the accelerator pedal. The resulting power at the wheels is compared with the desired power.

3.4.1   Simulation result

In Fig. 3.5 the wheel power for the controlled ZI powertrain with and without the flywheel is shown as a function of time. The desired wheel power is indicated by P_d. From Fig. 3.5 the following observations are made.

- The flywheel clearly acts as a peak-shaver: the response of the wheel power is immediate and relatively well persistent. The powertrain without the flywheel suffers from reluctance for increasing and decreasing power demands.
- The flywheel enlarges the rotating inertia of the powertrain. For maximal power this causes a difference between the desired and realized wheel power. This difference disappears as soon as a stationary vehicle speed is reached.
- The applied controller is not able to totally phase out (undesired) power oscillations. This is a subject of current investigations.

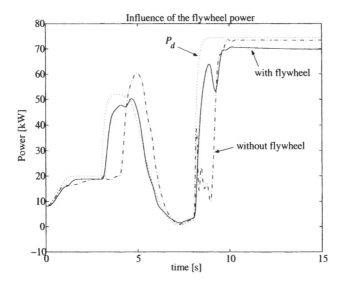

Fig. 3.5   Wheel power response of the CVT and ZI-CVT powertrains

3.4.2   Experimental results

In the test rig experiments the ZI transmission (i.e. all parts of the ZI powertrain between the engine and the wheels) is tested. The combustion engine is replaced by a controllable electric motor whereas a controllable electric brake generates the load of the powertrain. With this setup, a fairly realistic simulation of both the combustion torque and the wheel load is possible. However, the inertia of the electric motor is much larger than that of a combustion engine. Therefore, the results of dynamic experiments can differ significantly from those in a real car.

Three experiments are performed. First, the efficiency of the ZI transmission is determined as a function of the vehicle speed. The results are given in Fig. 3.6. It turns out that the efficiency is at most 2 per cent lower than that of the ZI transmission after removal of the flywheel. This degradation of the efficiency is due to spindle and air drag losses within the flywheel unit.

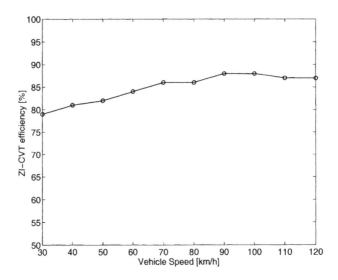

Fig. 3.6 Efficiency of the ZI-CVT

In the second experiment the power at the wheels after a kick down of the drive pedal was measured for the ZI transmission with and without the flywheel. The results are shown in the lower part of Fig. 3.7. These plots in fact give some information on how well the driveability problem is solved by the incorporation of the flywheel in the transmission. The upper part of this figure gives the difference of these powers as a function of time. It is seen that within half a second the wheel power for the transmission with the flywheel is about 35 kW larger than that for the transmission without the flywheel. After 1.2 seconds there is hardly any difference anymore. The fairly large oscillations for $t>1$ second are caused by the fact that the inertia of the electric motor in the test rig is much larger than that of a combustion engine. The electric motor is compensated for this higher inertia through a setpoint correction. This correction looses effectiveness during induced oscillations since phase-lags in the corrected setpoint become too high. This difference between the test rig and a real car also causes the oscillations in the third experiment around $t=10$ seconds in the right-upper plot of Fig. 3.8. Here, a virtual driver had to track a desired vehicle speed. This is shown in the upper-left plot together with the engine and flywheel speed. The lower-left plot in Fig. 3.8 illustrates that the E-line is tracked fairly well. As of the limited bandwidth of both the motor and CVT it is not possible to track this line exactly, especially not during large transients. The effect of this limited bandwidth is shown in the lower-right plot in Fig. 3.8. Here, the absolute relative difference between desired and realized wheel power, torque, and speed are shown. In fact, these are the 1-norms in equations (3.11) and (3.12). It can be seen from these graphs that according to these norms the tracking of the E-line succeeds relatively better than fulfilling the desired wheel power. The DC-offsets in the normed power error can be exemplified by the limited ZI-CVT effiency, *cf.* Fig. 3.6.

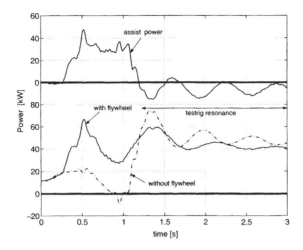

Fig. 3.7 Initial wheel response of the ZI-CVT

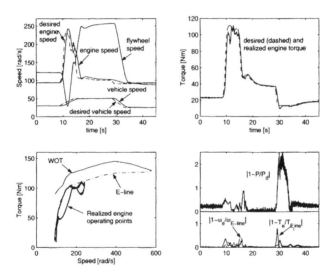

Fig. 3.8 E-line tracking during city driving

3.5   Conclusion

Regarding the results discussed in the previous section the following conclusions are drawn.

• Fuel consumption on an NEDC is expected to be improved by 15 per cent with respect to a 4-AT powertrain. This will be verified in the near future, using a test vehicle equipped with the ZI powertrain.

- The improvement of the driveability due to the flywheel unit is large. Power responses are immediate and persistent.
- These levels of power assist may lead to a preferably smaller flywheel inertia after judging the results of a (subjective) driveability assessment.
- Improvements of the powertrain controller should focus on the attenuation of unwanted oscillations.
- Improvements on the mechanical design should focus on an integrated design of the ZI transmission with less components, and volume and weight.

3.6   References

(1)   **Dorisen, H.T.** and **Höver, N.**, (1995), Antriebsschlupfregelung (ASR) – Ein beitrag zur aktiven fahrsicherheit, *Automobiltechnische Zeitschrift (ATZ)* **95**–4, 194–201.

(2)   **Druten, R.M. van**, *et al.*, (2000), Design and construction aspects of a zero inertia CVT for passenger cars, Proceedings of the *FISITA World Automotive Congress*, Seoul.

(3)   **Ide, T., Udagawa, A.**, and **Kataoka, R.**, (1994), A dynamic response analysis of a vehicle with a metal V-belt CVT, *Proc. AVEC '94*, **1**, 230–235, Tsukuba.

(4)   **McRuer, D.** and **Krendel E.**, (1957), Dynamic response of human operators, *Columbus Wright Air Development Center.*

(5)   **Serrarens, A.F.A.** and **Veldpaus, F.E.**, (2000), New concepts for control of power transients in flywheel assisted drivelines with a CVT, Proceedings of the *FISITA World Automotive Congress*, Seoul.

(6)   **Pfiffner, R.**, (1999), Steuerstrategien von CVT-Getrieben: Ubersicht und Ausblick, *VDI Berichte Nr. 1467*, 313–331.

(7)   **Spijk, G. van** and **Veenhuizen, B.**, (1999), An upshift in CVT efficiency, *VDI Berichte Nr. 1393.*

(8)   **Vroemen, B.G.** and **Veldpaus, F.E.**, (2000), Control of a CVT in a flywheel assisted driveline, Proceedings of the *FISITA World Automotive Congress*, Seoul.

(9)   Ford 1997. http://www.uscar.org/techno/hyb_ford.htm

(10)   Honda 1999. http://www.evworld.com/reports/hondavv.html

4

Performance of an Integrated Engine-CVT Control, Considering Powertrain Loss and CVT Response Lag

T Kim and **H Kim**

Abstract

In this chapter, an integrated engine-CVT control algorithm is suggested by considering the powertrain loss and the inertia torque due to the CVT ratio change during the transient state. In addition, compensation algorithms to reduce the effect of the CVT ratio response lag on the drive torque are presented. Experimental results show that the optimal engine speed compensation algorithm gives better engine operation around the optimal operation line, compared to the optimal torque compensation, while showing nearly the same acceleration response. The performance of the proposed integrated engine-CVT control algorithms is compared with those of conventional CVT control, and it is found that optimal engine operation can be achieved by using integrated control during acceleration, and improved fuel economy can be expected while also satisfying the driver's demands.

4.1 Introduction

If engine operation is maintained in the minimum fuel consumption region under all driving conditions, a significant improvement in fuel economy can be expected. For a conventional powertrain which consists of an internal combustion engine and an automatic transmission, power output is determined by the throttle valve opening (TVO), which is operated directly by the accelerator pedal. The automatic transmission controls the speed ratio depending on the TVO and the vehicle velocity. Since the engine operation point is determined by the TVO and the transmission ratio, it is impossible to maintain an optimal engine operation at all times in the minimum fuel consumption region. In order to achieve minimum fuel consumption relative to various levels of desired drive torque, both the engine speed and the engine torque should be controlled simultaneously, which requires an integrated engine-transmission control. Since an automatic transmission which has a stepped gear ratio cannot provide the optimal engine operation independent of the vehicle speed, a continuously variable transmission (CVT) is required for integrated engine-transmission control.

An engine-CVT integrated control algorithm has been suggested by Takiyama (**1**). He developed an algorithm to control the vehicle velocity from the difference between the desired and actual velocity and the CVT ratio from the difference between the desired and actual engine speed. However, he neglected transient characteristics of the powertrain, which resulted in poor performance when the magnitude of the vehicle acceleration changed. Recently, Takiyama (**2**) investigated an integrated engine-CVT control algorithm with combined air–fuel ratio control to realize better fuel economy. In the integrated engine-CVT control, the target engine torque should be determined by considering the powertrain loss. Sakaguchi (**3**) suggested an algorithm from the viewpoint of minimizing powertrain loss.

In a CVT powertrain, the drive torque is a product of the engine torque and the CVT ratio. In a situation where a large step-like increase in the drive torque is demanded, such as in kickdown maneuvers when accelerating rapidly, the response lag of the engine torque and that of the CVT ratio change translate directly into a drive torque response delay. Yasuoka (**4**) developed an algorithm to obtain the demanded drive torque for optimum fuel economy. In his algorithm, he used the engine torque to compensate for the drive torque response delay caused by the CVT response lag. Yasuoka calculated the target torque by assuming that the accelerator pedal travel represents the demanded drive torque and used the target gear ratio as the CVT ratio. In previous work by the authors (**5**), Kim *et al.* Suggested a fuzzy logic based integrated control algorithm considering the CVT shift dynamics.

In this chapter, an integrated engine-CVT control algorithm is suggested to improve the driveability of the CVT vehicle by considering the powertrain loss and inertia torque due to the CVT ratio change. In addition, algorithms to compensate the drive torque response delay due to the CVT response lag are presented. Experiments are carried out to investigate performance of the integrated engine-CVT control algorithm suggested in this study.

4.2   Optimal engine operation

The purpose of the integrated engine-CVT control is to realize the optimal engine operation for minimum fuel consumption while also satisfying the driver's demand. For optimal engine operation, the engine should be operated on the optimal operating line (OOL). In Fig. 4.1, an OOL for minimum fuel consumption is shown on the engine characteristic map with TVO and iso-power curves. The OOL for the minimum fuel consumption can be determined from BSFC contours and iso-power curves. The optimal engine operation point is defined as the point where the optimal engine power curve crosses with the OOL. Minimum fuel consumption can be achieved by operating the engine on the optimal operation point by simultaneous TVO and CVT ratio control, in other words, an integrated control.

In order to perform the integrated engine-CVT control, the optimal engine operation point needs to be determined at first for a given driving condition. The optimal engine operation point is determined from the OOL and the optimal engine power curve as described earlier. The optimal engine power can be obtained from the power required to drive the vehicle.

The power required for the vehicle $P_{v\,desired}$ is calculated by the following equation:

$$P_{v\ desired} = (F_L + M\dot{V}_{desired})V_{desired} \tag{4.1}$$

where F_L is the road load, M is the vehicle mass, $\dot{V}_{desired}$ is the desired vehicle acceleration, and $V_{desired}$ is the desired velocity. In an ideal situation without powertrain loss, the optimal engine power is obtained as:

$$P_{e\ optiaml} = P_{v\ desired} \tag{4.2}$$

From the optimal engine power and the OOL, the optimal engine torque, $T_{e\ optimal}$ and the optimal engine speed, $\omega_{e\ optimal}$, can be obtained from Fig. 4.1. In addition, the optimal TVO to generate $T_{e\ optimal}$ is also determined from the optimal operation point. The optimal CVT ratio, $i_{optimal}$, can be determined from $\omega_{e\ optimal}$ and the actual vehicle velocity, V, as follows:

$$i_{optiaml} = \frac{\omega_{e\ optimal}}{N_{frg}\omega_v} \tag{4.3}$$

where N_{frg} is the final reduction gear ratio and ω_v is the vehicle velocity. The above algorithm can be applied for steady state driving situation without powertrain loss. However, in the real world, there exists a powertrain loss. Furthermore, the effect of the inertia torque and the CVT response lag should be considered during the transient state when the CVT ratio changes. In the next section, an engine-CVT integrated control algorithm will be discussed by considering the powertain loss, inertia torque, and CVT ratio change response lag.

4.3   Powertrain loss compensation

Among the sources of powetrain loss of a CVT vehicle, CVT loss is considered to be a major contributing factor. The CVT loss consists of a transmission loss and a hydraulic loss. The transmission loss occurs due to the slippage between the belt and the pulley (6). Considering the transmission loss, the optimal engine power is expressed as:

$$P_{e\ optiaml} = \frac{P_{v\ desired}}{\eta_{cvt}} \tag{4.4}$$

where η_{cvt} is the transmission efficiency. The transmission loss due to the belt slippage is included in η_{cvt}. CVT hydraulic loss, P_{hyd}, is determined from the line pressure, P_{line}, and the pump flow rate, Q_{pump}, as:

$$P_{hyd} = P_{line}Q_{pump} \tag{4.5}$$

Therefore, the optimal engine power is modified as:

$$P_{e\ optiaml} = \frac{P_{v\ desired}}{\eta_{cvt}} + P_{hyd} \tag{4.6}$$

In equation (4.6), P_{hyd} is determined from the line pressure and the engine speed. Since the line pressure depends the engine torque and the CVT ratio that can be obtained from the optimal engine power, it is impossible to determine the engine torque and the CVT ratio with unknown optimal engine power. Therefore, in this study, P_{hyd} is estimated by the following procedure. Firstly, the engine torque, engine speed, and the CVT ratio are obtained at the optimal operation point from equation (4.2) by neglecting the powertrain loss. Using these values, P_{hyd} can be calculated. With known values of P_{hyd}, we can calculate the modified optimal engine power, $P_{e\ optimal}$ by equation (4.6). From the modified $P_{e\ optimal}$ and the OOL, a new optimal operation point is determined and modified values of $\omega_{e\ optimal}$, $T_{e\ optimal}$, $i_{optimal}$ are obtained at this point. Now, we can calculate new P_{hyd} from modified $\omega_{e\ optimal}$, $T_{e\ optimal}$, $i_{optimal}$. The more this procedure is repeated, the more exact value of the powertrain loss, in other words, the optimal engine power can be obtained. In this study, the powertrain loss is determined by carrying out this procedure two times considering the computation time for the integrated control.

4.4   Inertia torque compensation

The effect of the inertia torque during the transient state is investigated in this section. Dynamic equation of the CVT vehicle is represented as:

$$\dot{\omega}_v = \frac{N_{frg}i(T_e - T_{eloss}) - T_{load} - N_{frg}i\dfrac{di}{dt}J_{eq}\omega_v}{J_v + J_{eq}N_{frg}^2 i^2} \tag{4.7}$$

where $\dot{\omega}_v$ is the vehicle acceleration, T_e is the engine torque, T_{eloss} is the engine torque loss, T_{load} is the road load, i is the actual CVT ratio, J_v is the vehicle inertia, J_{eq} is the equivalent inertia viewed from the engine side. The CVT ratio change rate is di/dt. The last term in the right side of equation (4.7) is the inertia torque due to the CVT ratio change, di/dt. di/dt has a positive value during downshift when the speed ratio increases and has a negative value during upshift. Therefore, the inertia torque when the speed ratio increases contributes to reducing the acceleration torque, which results in hesitation of the vehicle response. In this study, the power due to the inertia torque is defined as:

$$P_{inertia} = N_{frg}i\frac{di}{dt}J_{eq}\omega_v\omega_e \tag{4.8}$$

Compensating the power due to the inertia torque, the optimal engine power is modified as:

$$P_{e\ optiaml} = \frac{P_{v\ desired}}{\eta_{cvt}} + P_{hyd} + P_{inetia} \tag{4.9}$$

4.5   CVT ratio change response lag compensation

Vehicle drive torque is a product of the engine torque and the CVT ratio. In general, 90 per cent of the engine torque response is obtained within several hundred milliseconds, whereas that of a belt drive CVT takes around one second. Therefore, the CVT ratio change response lag has a large effect on the response delay of the drive torque during acceleration such as a kickdown maneuver. Therefore, in order to improve the driveability, a remedy to reduce the effect of the CVT response lag on the drive torque should be provided.

4.5.1   Optimal engine torque compensation

In order to compensate the drive torque response lag due to the CVT ratio change response lag, the following algorithm is suggested:

$$T_{e\ compensated} = T_{e\ optimal} \frac{i_{optimal}}{i} \tag{4.10}$$

where $T_{e\ compensated}$ is the compensated engine torque and $T_{e\ optimal}$ is the optimal engine torque which is determined from Fig. 4.1. During acceleration such as in a kickdown maneuver, i follows $i_{optimal}$ with a smaller value than $i_{optimal}$, which results in a larger $T_{e\ compensated}$ than $T_{e\ optimal}$ from equation (4.10).

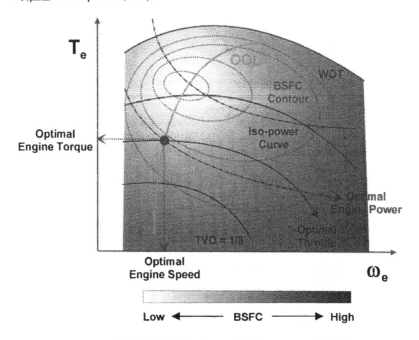

Fig. 4.1 Engine characteristic curves and OOL

4.5.2   Optimal engine speed compensation

Another way to increase the drive torque during the acceleration is to increase the CVT ratio. If the optimal CVT ratio, $i_{optimal}$ which is larger than the optimal ratio is created, the larger actual ratio, i will be generated. Since drive torque is a product of the engine torque and the actual speed ratio, an increased drive torque can be obtained. One way to obtain a larger $i_{optimal}$ is to increase the optimal engine speed, $\omega_{e\ optimal}$ since the speed ratio i is determined from the ratio between the optimal engine speed and the actual vehicle speed. Therefore, in this study, an algorithm to increase $\omega_{e\ optimal}$ is presented by using $i_{optimal}$ and i:

$$\omega_{e\ compensated} = \omega_{e\ optimal} \frac{i_{optimal}}{i} \tag{4.11}$$

where $\omega_{e\ compensated}$ is the compensated engine speed. $\omega_{e\ optimal}$ is determined from optimal operation point in Fig. 4.1.

In Fig. 4.2, a block diagram of the integrated engine-CVT control algorithm suggested in this study is shown. The integrated control algorithm consists of (1) the optimal engine power calculation, and (2) the optimal TVO and CVT ratio determination. The optimal engine power is calculated from the desired vehicle power by considering powertrain loss and inertia torque due to the CVT ratio change. From the optimal engine power and the OOL, the optimal engine operation point is determined where the optimal engine torque, speed can be obtained. The optimal TVO is determined from the optimal operation point. The optimal CVT ratio is determined by the engine speed compensation algorithm suggested in this study considering the CVT ratio response lag.

4.6   Experimental results and discussion

Figure 4.3 shows a picture of CVT bench tester used in this study. A three cylinder 800 cc engine ① is used with a "drive by wire" throttle control system. A CVT ② equipped with a stepping motor drive electronic ratio control system is used. A flywheel ④ is used to simulate a 900kg vehicle mass. Eddy current dynamometer ⑤ is used to simulate a road load.

Figures 4.4–4.7 show experimental results for the integrated engine-CVT control. In Fig. 4.4, performance of the integrated control with powertrain loss compensation are compared with the results without compensation. As shown in Fig. 4.4, the engine power (a) and the TVO (b) with the powertrain loss compensation are larger than those without the compensation. Consequently, the drive torque (c) and the velocity (d) with the compensation follow the reference more closely than those without the compensation. The time delay of the drive torque (c) in the beginning is due to the response lag of the CVT ratio change.

Figure 4.5 shows experimental results for the integrated engine-CVT control with inertia torque compensation. The performance of the CVT vehicle with inertia torque compensation is compared with the results without compensation. In the experiments, powertrain loss compensation is included. It is seen from Fig. 4.5 that the engine torque (a) with the inertia torque compensation shows a larger value than the engine torque without compensation. The drive torque (c) also increased corresponding to the increased engine torque. Consequently, the vehicle velocity with the inertia torque compensation follows the desired velocity more closely than the vehicle velocity without compensation.

In Fig. 4.6, experimental results for the integrated engine-CVT control with CVT ratio response lag compensation by the torque compensation algorithm, are compared with the results without compensation. In the experiments, the powertrain loss and the inertia torque compensations are included for both cases. As shown in Fig. 4.6, the engine torque (a) with the CVT response lag compensation increases during acceleration by equation (4.10). This is because the actual CVT ratio (b) shows a smaller value than the optimal ratio as expected. The increased engine torque results in an increased drive torque (c). Therefore, the vehicle velocity with compensation follows the reference velocity more closely than the vehicle velocity without compensation. In Fig. 4.6(e), the engine operation trajectory for the integrated engine-CVT control with the CVT response lag is plotted during acceleration with the OOL. It is observed from Fig. 4.6(e) that the engine operation is carried out above the OOL since the engine speed cannot change rapidly compared with the engine torque change. The purpose of

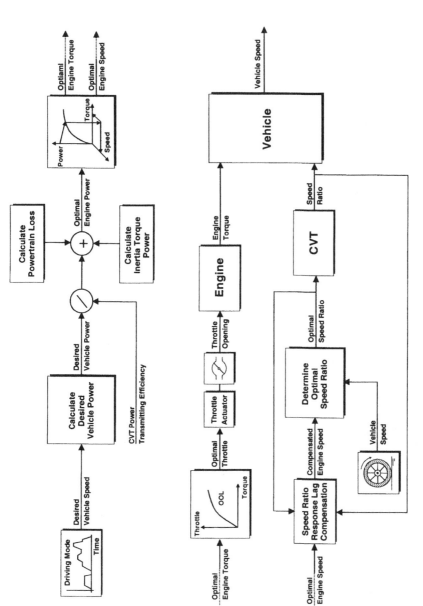

Fig. 4.2 Block diagram of integrated engine-CVT control algorithm

integrating the engine-CVT control is to achieve the optimal engine operation while also satisfying the driver's demand. From Fig. 4.6, we can see that even if the acceleration performance is improved by the optimal engine torque compensation, the engine operation is performed outside the OOL during acceleration.

Fig. 4.3 CVT bench tester

In Fig. 4.7, experimental results for the integrated engine-CVT control with the CVT response lag compensation by the speed compensation algorithm, are compared with the results without compensation. As shown in Fig. 4.7, the engine torque (a) with compensation does not change much from the torque without compensation since the optimal engine speed is only increased for the CVT ratio response lag compensation. The optimal CVT ratio (b) changes showing an increased value compared with the CVT ratio by the engine torque compensation (Fig. 4.6(b)). The actual CVT ratio with compensation is larger than the ratio without compensation. Therefore, the drive torque (c) with compensation is increased owing to the increased CVT ratio, which results in improved response of the vehicle velocity (d). The engine operation trajectory (e) during acceleration is plotted with the OOL. Compared with the engine operation trajectory with optimal engine torque compensation (Fig. 4.6(e)), it is seen that improved engine operation can be achieved by the engine speed compensation algorithm. Therefore, we can say that for the CVT ratio response lag compensation, the optimal engine speed compensation algorithm gives better engine operation around the OOL than the engine torque compensation while showing almost the same vehicle velocity response.

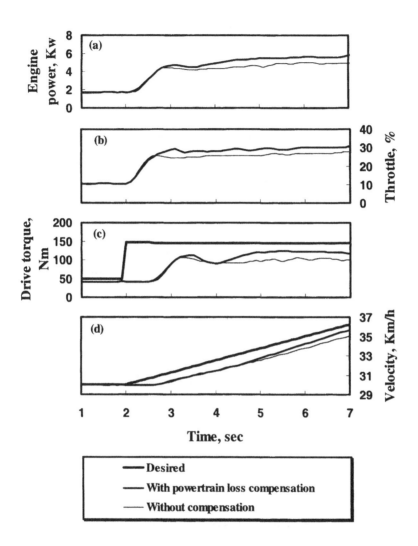

Fig. 4.4 Experimental results for powertrain loss compensation

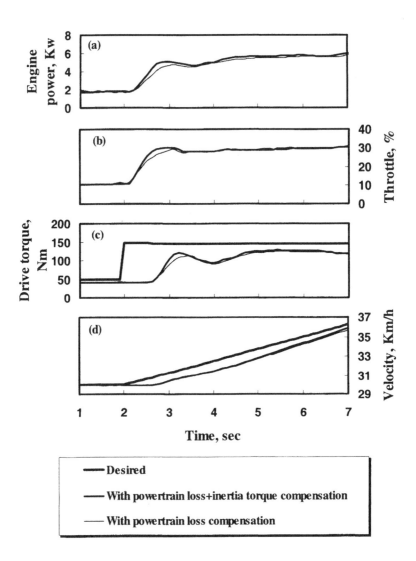

Fig. 4.5 Experimental results for inertia torque compensation

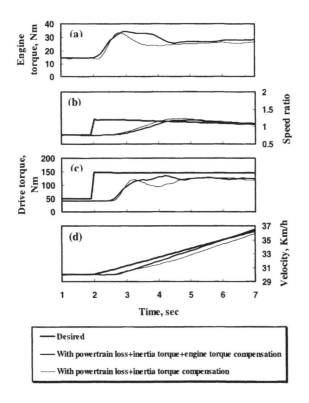

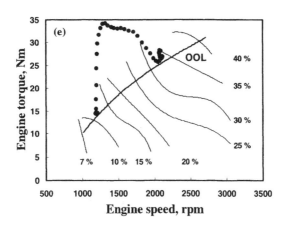

Fig. 4.6 Response for engine torque compensation

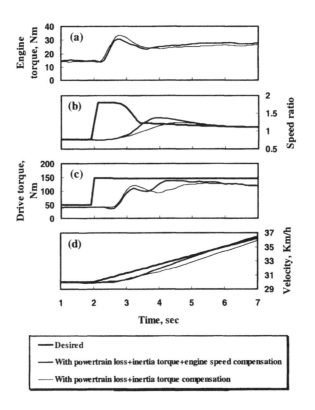

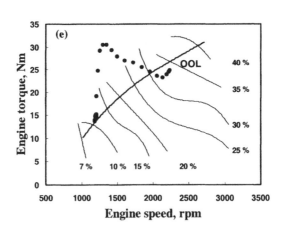

Fig. 4.7 Response for engine speed compensation

4.7   Comparison of integrated engine-CVT control algorithm with conventional CVT control

In order to evaluate the integrated engine-CVT control algorithm developed in this study, vehicle performance using the integrated control is compared with those of the conventional CVT control.

Figure 4.8 shows comparison of the experimental results in acceleration. In the experiments, the engine speed compensation algorithm was used for the integrated control. The engine torque (a) by the integrated control increases showing lower values than the engine torque by conventional CVT control to achieve the optimal engine operation. The CVT ratio (b) by the integrated control downshifts showing a lower speed ratio in the beginning stage of the acceleration than the ratio by conventional CVT control, according to the engine speed compensation algorithm. The drive torque (c) changes according to the response of the CVT ratio and the engine torque since the drive torque is a product of the engine torque and the CVT ratio. The drive torque by the integrated control shows a lower value than that of the conventional CVT control in the initial state of the acceleration, and shows a higher value in the remaining region to achieve the optimal engine operation by the engine speed compensation algorithm. It is seen from Fig. 4.8(d) that the vehicle velocity by the integrated control shows a slightly slower response in the initial stage of the acceleration, but shows almost the same response with the conventional CVT control. In Fig. 4.8(e), the engine operation trajectory is compared with the OOL. The engine operation point moves above the OOL in the initial stage of the acceleration for both the integrated and conventional control since the engine torque increases more rapidly than the engine speed. However, it is seen that the engine operation by the integrated control is carried out near the OOL during acceleration period compared with those of the conventional CVT control.

It is seen from the experimental results that the integrated engine-CVT control algorithm suggested in this study is able to provide an improved engine operation near the OOL in acceleration while satisfying the driver's demands. It is expected that the improved engine operation on the OOL for the minimum fuel consumption is able to achieve an improvement of the fuel economy.

4.8   Conclusion

An integrated engine-CVT control algorithm is suggested by considering the powertrain loss, the inertia torque due to the CVT ratio change. In determining optimal engine power, the powertrain loss was obtained by iteration procedure. In addition, in order to compensate for the drive torque time delay due to the CVT ratio response lag, engine torque and engine speed compensation algorithms are suggested. It is found from the experimental results that the optimal engine speed compensation algorithm gives better engine operation around the optimal operation line compared to the optimal torque compensation while showing nearly the same acceleration response. The performance of the proposed integrated engine-CVT control algorithm is compared with those of conventional CVT control. It is found that better optimal engine operation can be achieved by using integrated control during acceleration, while also satisfying the driver's demands. It is expected that the improved engine operation on the OOL

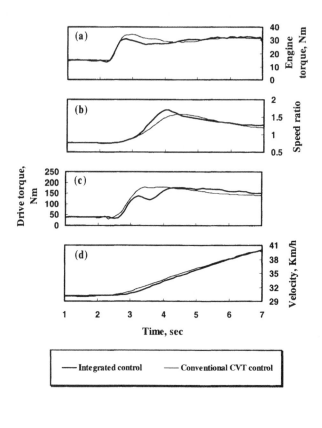

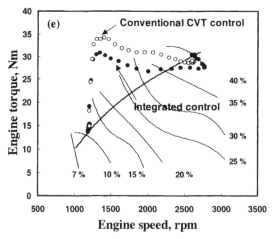

Fig. 4.8 Comparison of integrated control with conventional CVT control

for the minimum fuel consumption is able to provide an improvement of the fuel economy by the integrated control.

4.9   References

(1)   **Takiyama, T.** and **Morita, S.,** (1991), Comparison study for improvement of fuel consumption of ENGINE-CVT control, *Proc. JSME*, **58**, No. 547, 954–958.

(2)   **Takiyama, T.,** (1999), Engine-CVT consolidated control for variable A/F, *Proc. Int. Conf. on Continuously Variable Transmission*, 36–41.

(3)   **Sakaguchi, S., Kimura, E.,** and **Yamamoto, K.,** (1999), Development of an engine-CVT integrated control system, *SAE Paper* 1999-01-9574.

(4)   **Yasuoka, M., Uchida, M., Katakura, S.,** and **Yoshino, T.,** (1999), An integrated control algorithm for an SI engine and a CVT, *SAE Paper* 1999-01-0752.

(5)   **Kim, T.C.** and **Kim, H.S.,** (1997), Study on engine-CVT consolidated control (I) – development of consolidated control algorithm, *Trans. of KSAE*, **5**, No. 5, 86–96.

(6)   **Guebeli, M., Micklem, J.D.,** and **Rurrows, C.R.,** (1992), Maximum transmission efficiency of a steel belt continuously variable transmission, *Proc. of ASME Int. Power Transmission and Gearing Conf.*, **43**–1, 329–334.

5

Shifting Dynamics of Metal Pushing V-Belt – Rapid Speed Ratio Variations

G Carbone, L Mangialardi, and **G Mantriota**

Abstract

This work concerns with the investigation of transient dynamics of metal pushing V-belt when the pitch radius changes quickly. The metal V-belt consists of wedge-shaped metal plates that are supported by a flexible metal band. Two typologies of metal V-belt exist; one with no clearance among the metal plates and the other with clearance existing among the plates. This chapter investigates the behaviour of the V-belt by taking into account the effects of clearance presence.

One dimensional continuous model of the belt is developed in this work. The model neglects, for what concerns the estimation of the relative velocity between pulley and belt, the effect of plates and pulleys strain motion with respect to rigid one and considers no friction between band and plates.

The power transmission is assured only if an active arc exists where the plates are pressed together and the clearance vanishes. In effect, compression forces have to arise among the metal plates in order to transmit torque. Conversely, on the idle arc the plates are separated and no longitudinal compression forces exist among the plates.

5.1 Introduction

A Continuously Variable Transmission (CVT) is a power transmission device whose speed ratio can be varied in a continuous manner. There are many kinds of CVTs having different characteristics and V-belt CVTs are the most applicable in practice. When compared with a torque converter, a V-belt CVT can be expected to have a higher transmission efficiency, a simpler mechanism and many other merits. It is widely used in agricultural vehicles, mini cars and motorcycles. Recently, metal pushing V-belt and metal chain CVTs have been developed for automotive applications. In this field CVT transmissions have another advantage as they allow to vary the speed ratio under load condition without the necessity to disengage the engine by means the friction clutch.

Nowadays more and more attention is placed on vehicles' comfort, fuel saving, and low environmental impact. To achieve this objective continuously variable transmission (CVT) could be helpful. CVT transmission can provide a better performance of vehicle especially for what concerns the fuel consumption, pollution, comfort, and driveability. But a powerful control of the CVT transmission is needed to achieve this goal. Today, these transmissions equip little and mid cars but it seems realistic the use of CVT on vehicles with greater power and in the energy production systems: applications in city buses (1), in waterpumping windmills and wind power systems windmills (2)–(4) were proposed. Moreover, theoretical investigation are being conducted on Infinitely Variable Transmission (IVT), whose main components are the CVT unit a fixed ratio gear and a planetary gear (5).

Several researchers have been studying CVT, but their works, generally are concerned with steady-state dynamics of V-belt CVT (6)–(10). Not many works regard the transient dynamics of CVT: experimental investigations were made by Ide *et al.* (11) which derived a differential equation to describe the trend of speed ratio against the time, Kanehara *et al.* (12), (13) characterized the shifting dynamics by means a suitable mean friction coefficient, which identifies the slip condition of the belt, i.e. the extension of the active arc on the pulley, and by using the shifting gradient parameter.

In automotive applications, the axial thrust on the half-pulleys are, in some operative conditions, too large, in order to avoid the possibility of global sliding between the belt and the pulley, with negative effects on the transmission's efficiency. This consideration suggests to further investigate the mechanics of CVT in transient condition with the aim to choose the better solutions and particularly the better axial thrust in order to improve the transmission's efficiency.

Theoretical investigation of transient dynamics of metal pushing V-belt are carried on in order to derive a relation between some macroscopic quantities easily measurable: axial thrust, torque transmitted and belt's tensions on the slack and tight side (14), (15).

This chapter investigates the behaviour of metal pushing V-belt by taking into account the effects of the clearance among the metal plates. One-dimensional continuous model of the belt is developed, it neglects, for what concerns the estimation of the relative velocity between pulley and belt, the effect of plates and pulleys deformations and considers no friction between the band and the plates.

Figure 5.1 shows the metal pushing V-belt structure – the steel wedge-shaped plates, supported by the steel band, are visible. Moreover, the belt's construction makes it very flexible so that the bending stiffness of the belt is neglected in this work.

The metal blocks are treated as rigid body in their motion. In effect we are primarily interested in rapid pitch radius changes, therefore the strain motion of plates can be neglected with respect to rigid one for the evaluation of relative velocity (sliding velocity) between belt and pulley needed to calculate the friction forces. In steady state operation this means the relative motions of the belt being purely tangential directed on the active arc and that the Euler's equation can be used on this arc. Gerbert (8) demonstrated this being true when the clearance between metal plates is not too small.

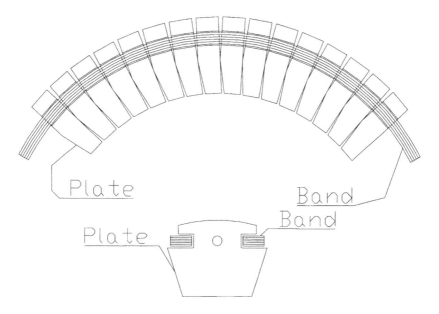

Fig. 5.1 Metal Pushing V-Belt

5.2 Metal belt's mechanics

First of all it is necessary to establish where the metal blocks are compressed and where not. In stationary condition, i.e. without speed ratio's changes, Gerbert (**8**) (Fig 5. 2) pointed out the existence, on the driver pulley, of both an idle arc, where the plates are separated, and an active arc with no clearance among the plates. Conversely, he underlined that no clearance exists among the metal blocks on the driven pulley. We consider this being true also in transient conditions, in effect on the driven pulley the blocks' speed has to exceed pulley's speed to guarantee, by means the friction forces, the power transmission. Hence, if the blocks were separated, the friction forces would force the blocks to decrease their velocity and the clearance to vanish immediately. The situation differs on driver pulley where both an idle arc and an active arc exist. On the idle arc the clearance among the plates differs from zero and no tangential slip exist between the blocks and the pulley, conversely on the active arc the metal plates are pressed each other in order to guarantee the power transmission. The section, which separates the idle arc from the active arc, is called the shock section because it represents the section where a sudden variation of tangential sliding speed of metal plates happens in order to quickly reduce to zero the clearance between plates: the compression force between two adjacent plates arises abruptly. The active arc with compression forces among the plates extents from the shock section to exit point, the tangential blocks' speed is lower than pulley's velocity in order to transmit power and this is consistent with continuity equation.

Obviously, on the pushing side of the belt, i.e. between the exit point of the driver pulley and the entry point of driven pulley, there is no clearance among the metal blocks. The pushing side of the belt is called the slack side of the belt. On the other side, i.e. between the exit point of driven pulley and the entry point of driver one, the clearance among plates differs from zero namely no compression forces exist among the plates – this is the tight side of the belt.

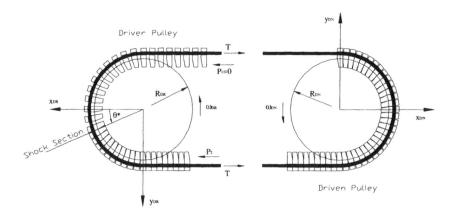

Fig. 5. 2 CVT Transmission

For what concerns the main hypotheses placed we suppose little tangential sliding velocity between the pulley and belt. This seems plausible because otherwise the transmission would work very bad with high power dissipation and lower durability of the belt. Therefore, the calculation of centrifugal forces acting on the belt will be made by considering the angular velocity of each belt's element equal to pulley's rotating speed. Moreover, we neglect all the other inertia belt's forces as the tangential inertia belt's force and the one related directly to the second derivative of the pitch radius R.

The radial thickness of the belt is neglected and the number of metal plates is supposed very high so that the thickness of each plate could be considered sufficiently small. Therefore, a one-dimensional continuous model of the belt is developed in the work.

For what concerns the interaction between the band and the metal plates and the tension of the band, Gerbert (**8**) assumed the band's tension being constant because he supposed that the band's stiffness was much lower that the plate's one, moreover Kim (**9**) assumed the band tension to be constant because he neglected the friction forces between band and blocks. Sun (**10**) underlined that the friction coefficient between band and blocks is about $\mu_b = 0.01$. In effect we can assert that the band behaves like a flat belt so that the Euler's equation can be used, namely, by neglecting the centrifugal forces on the band, $\dfrac{T_1}{T_2} = e^{\mu_b \alpha}$, where α is the slipping angle of the band. Now let assume $\alpha = \pi$ and $\mu_b = 0.01$, from the previous equation derives $\dfrac{T_1}{T_2} = 1.03$, namely the percentage difference between T_1 and T_2 is limited to 3 per cent. The last result means that the variations of band's tension can be neglected.

Moreover, together with the previous assumptions, in this work the bending stiffness of the belt and the strain motion of the belt and pulleys are neglected as it was highlighted previously.

We have already explained that on the driver pulley there are an idle arc with clearance among the metal plates, which have no tangential sliding velocity, and an active arc where the plates are pressed together in order to allow the power transmission. The two arcs are separated by the shock section, which doesn't exist on the driven pulley where the plates are pressed each other along the whole arc of contact and no clearance exist among the plates. Owning to this difference the dynamical behaviour of the belt is not the same on each pulley. Therefore, in order to describe the dynamics of the belt two different approaches are needed for the driver pulley and the driven one.

5.3 Dynamical equation where the plates are pressed together

Where the plates are pressed and longitudinal compression forces exist the clearance among the plates vanishes. The set of equations which describes the dynamics of the belt is obtained by applying the continuity equation, and the momentum equations to a generic element of the belt.

Let consider the generic element of the belt whose length is $ds = Rd\theta$, since the strain motion of the belt is neglected and no clearance among the plates exist, the length ds must be constant during the motion, hence $(d\dot{s}) = 0$ and by deriving the previous relation we obtain

$$\frac{\dot{R}}{R} + \frac{(\dot{d\theta})}{d\theta} = 0$$

Now the problem is to find an appropriate expression for the term

$$\frac{(\dot{d\theta})}{d\theta} = \frac{1}{d\theta}\frac{d(d\theta)}{dt} = \frac{1}{d\theta}\frac{d\theta(t+dt)-d\theta(t)}{dt}$$

By observing Fig. 5. 3, it follows that:

$$d\theta(t+dt) = \theta + d\theta(t) + \dot{\theta}(\theta+d\theta,t)dt - \left(\theta + \dot{\theta}(\theta,t)dt\right) = d\theta(t) + \frac{\partial\dot{\theta}}{\partial\theta}d\theta dt$$

therefore

$$\frac{(\dot{d\theta})}{d\theta} = \frac{\partial\dot{\theta}}{\partial\theta}$$

where $\dot{\theta}$ is the angular velocity of the belt's element. Now by introducing the sliding angular velocity $\omega_s = \dot{\theta} - \omega$, where ω is the pulley's rotating speed not depending on the angular co-ordinate θ, the previous relation becomes: $\frac{\dot{R}}{R} + \frac{\partial\omega_s}{\partial\theta} = 0$. The last equation can be rearranged giving the following congruency equation which relates the radial velocity of the belt, namely, the pitch radius variations of the belt, with the tangential sliding velocity of the belt:

$$\frac{\partial\,s}{\partial\theta} + w = 0 \qquad\qquad \text{Congruency Equation} \qquad\qquad (5.1)$$

where the dimensionless parameters are the sliding coefficient $s = \dfrac{\omega_s}{\omega}$ and the dimensionless radial velocity $w = \dfrac{\dot{R}}{\omega R}$ (14), (15).

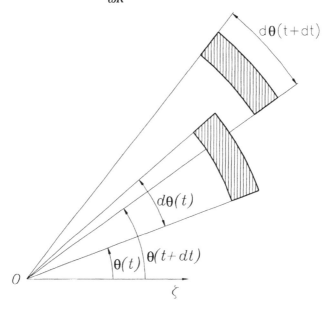

Fig. 5. 3 Displacement of a belt's element

Now the momentum equations may be written by observing the Fig. 5.4 where the forces acting on the belt's element are showed. The equilibrium of the belt along the tangential and radial direction requires:

$$\frac{1}{F - \sigma \omega^2 R^2} \frac{\partial (F - \sigma \omega^2 R^2)}{\partial \theta} = \frac{\mu \cos \beta_s \sin \gamma}{\sin \beta - \mu \cos \beta_s \cos \gamma}$$

and

$$p = \frac{F - \sigma R^2 \omega^2}{2R(\sin \beta - \mu \cos \beta_s \cos \gamma)}$$

where p is the linear pressure on the belt and $F = T - P$ is the net belt's tension defined as the difference between the constant band tension T and the compressive force P acting on the metal plate.

Let us introduce the following two quantities: $\theta *$ and $F*$. $\theta *$ is the angular co-ordinate of the shock section when we refer to the driver pulley, while it represents the angular co-ordinate of the actual or imaginary point with null sliding angular velocity for the driven pulley where the plates are pressed together along the contact arc. $F*$ is the net belt tension in the point $\theta = \theta *$; let observe that, for the driver pulley, this value of net belt tension refers to a point located shortly after the shock section, since the net belt tension is not continuous in the shock section.

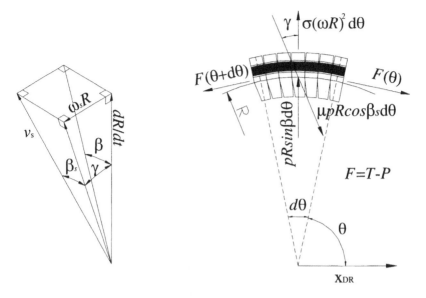

Fig. 5.4 Kinematic and dynamical quantities

Now by introducing the following dimensionless quantities $\kappa = \dfrac{F - \sigma R^2 \omega^2}{F* - \sigma R^2 \omega^2}$ and the dimensionless linear pressure $\tilde{p} = \dfrac{Rp}{F* - \sigma \omega^2 R^2}$, the previous relations can be rewritten as:

$$\frac{\partial \kappa}{\kappa \partial \theta} = \frac{\mu \cos \beta_s \sin \gamma}{\sin \beta - \mu \cos \beta_s \cos \gamma} \qquad \text{First Momentum Equation} \qquad (5.2)$$

$$\tilde{p} = \frac{\kappa}{2(\sin \beta - \mu \cos \beta_s \cos \gamma)} \qquad \text{Second Momentum Equation} \qquad (5.3)$$

Other two relations are needed which correlate the sliding coefficient s with the sliding angle γ and the pulley's half-opening angle β_s in the sliding plane with the sliding angle γ (see Fig. 5.4).

$$s = w \tan \gamma \qquad \text{Kinematic Equation} \qquad (5.4)$$

$$\tan \beta_s = \tan \beta \cos \gamma \qquad \text{Geometric Equation} \qquad (5.5)$$

where β is the half-opening angle of the pulley's groove.

5.4 Clearance between plates and kinematic strain of the belt

Indicate (Fig. 5.5) with the symbols da, and db the thickness of metal plates and the distance between the forward and backward faces of two adjacent plates respectively, measured along the circle with radius R. We place $\varepsilon = \dfrac{db}{da}$ and $ds = da + db$ where s represents the curvilinear co-ordinate and ds the distance between two adjacent blocks. Obviously $ds = (1 + \varepsilon)da$. We call ε the kinematical strain of the belt.

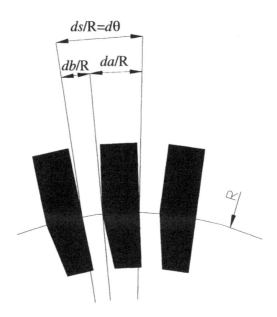

Fig. 5.5 Clearance between plates

5.5 Congruency equation where the plates are separated – driver pulley

On the driver pulley an active arc and an idle arc exist, they are separated by the shock section where the clearance between plates goes zero suddenly. The idle arc extends from the entry points to the shock section. On the idle arc the tangential velocity of plates equals the pulley's speed so that friction forces cannot have components in this direction – the dimensionless tension cannot change on the idle arc where the blocks move only outward or inward in the radial direction.

Since the clearance among the plates is not zero on the idle arc, the distance among the plates will vary during the motion, namely the length of a generic belt's element is not constant in the time. Therefore in order to write the congruency equation the derivative of the length ds has to be taken into account. By remembering that $ds = (1 + \varepsilon)da$ the following relation holds

$\dfrac{(\dot{ds})}{ds} = \dfrac{\dot{\varepsilon}}{1+\varepsilon}$, therefore the congruency equation becomes $\dfrac{\dot{\varepsilon}}{1+\varepsilon} = \dfrac{\dot{R}}{R} + \dfrac{\partial \omega_s}{\partial \theta}$. But on the idle

arc $\omega_s = 0$, namely $\dot{\theta} = \omega$, hence the previous equation simplifies:

$$\frac{1}{1+\varepsilon}\left(\frac{\partial \varepsilon}{\partial t} + \omega \frac{\partial \varepsilon}{\partial \theta} \right) = \frac{\dot{R}}{R} \tag{5.6}$$

The second hand side of the previous equation 5.6 depends only from the time t owning to no transversal deformation of plates, i.e. the radial position of blocks is uniform along the belt.

This equation can be rewritten as:

$$\frac{\partial \left(\ln \left| \frac{1+\varepsilon}{R} \right| \right)}{\partial t} + \omega \frac{\partial \left(\ln \left| \frac{1+\varepsilon}{R} \right| \right)}{\partial \theta} = 0 ,$$

therefore the general solution of equation 5.6 is:

$$\varepsilon = R\, f(\theta - \omega t) - 1 \tag{5.7}$$

where f is a generic function of the argument $\theta - \omega t$

We can see that the solution (equation 5.7) is very similar to the wave equation's solution with angular propagation speed equals to ω.

If we suppose ω sufficiently high, it is possible to neglect the variation of kinematical strain with angular co-ordinate θ and consider the kinematical strain ε uniform on the idle arc, namely $\dfrac{\partial \varepsilon}{\partial \theta} = 0$, and equal to its value ε^* shortly before the shock section.

5.6 Tension and linear pressure where the plates are separated – driver pulley

On the idle arc the plates are separated and this requires no compression forces acting on the metal plates, hence the net belt tension equals the tension T of the band and under the hypotheses placed it doesn't change along the contact arc. Since $F^* = T - P^*$, where P^* is the compressive forces acting on the metal plates shortly after the shock section, we could evaluate the value κ_{idle} of dimensionless tension κ on the idle arc:

$$\kappa_{idle} = \frac{T - \sigma\omega^2 R^2}{F^* - \sigma\omega^2 R^2} = 1 + \frac{P^*}{F^* - \sigma\omega^2 R^2} > 1 .$$

Since the values of pressure acting on the belt is necessary in order to evaluate the axial thrust needed to equilibrate the pulley, the expression of the linear pressure on the idle arc has to be found:

$$p_{idle} = \frac{T - \sigma_{idle} R^2 \omega^2}{2R(\sin\beta \mp \mu \cos\beta)}$$

where $\sigma_{idle} = \sigma \dfrac{1}{1 + \varepsilon_{idle}}$ is the linear belt mass on the idle arc and ε_{idle} is the kinematical strain on the idle arc. The minus sign refers to positive value of $\dot{R}$ while the positive sign refers to $\dot{R} < 0$. Naturally, if the kinematic strain of the belt could be considered uniform on the idle arc and equal to ε^*, both p_{idle} and σ_{idle} are uniform too. On the idle arc the dimensionless pressure assumes the following value:

$$\tilde{p}_{idle} = \frac{1}{2(\sin\beta \mp \mu \cos\beta)}\left(\kappa_{idle} + \frac{\varepsilon_{idle}}{1 + \varepsilon_{idle}} \frac{\sigma\omega^2 R^2}{F^* - \sigma\omega^2 R^2} \right)$$

5.7 Study of motion in the shock section – driver pulley

Now it is necessary to determinate the relation between the dimensionless slip factor s^* and the kinematical strain ε^* in the shock section. By applying the congruency (continuity) equation to a infinitesimal control volume of the belt standing across the shock section and moving at the same speed of this section (Fig. 5.6) we conclude that:

$$\left(\omega R - \frac{d\theta^*}{dt} R \right) \frac{\sigma}{1 + \varepsilon^*} = \left[(1 + s^*)\omega R - \frac{d\theta^*}{dt} R \right]\sigma$$

and, since $\left| \dfrac{1}{\omega} \dfrac{d\theta^*}{dt} \right| \ll 1$, $\left| s^* \right| \ll 1$ and $\left| \varepsilon^* \right| \ll 1$, it follows:

$$s^* + \varepsilon^* = 0 \qquad\qquad\qquad (5.8)$$

Because ε^* is always positive, the previous equation shows s^* being always negative. This result is evident since the belt's speed decreases, once passed the shock section, to generate negative tangential relative velocity and permit the transmission of power due to friction forces.

Owning to no tangential slip between the plates and the pulley on the idle arc, the sliding angle γ is null if $\dot{R} > 0$ and equals to π if $\dot{R} < 0$. On the other side in correspondence of the shock section there is a suddenly variation of the plates' speed so that the angle γ^* (shortly after the shock section) is negative for $\dot{R} > 0$ and greater than π for $\dot{R} < 0$.

By applying the momentum equation to the same control volume used before (Fig. 5.6), it is possible to evaluate the magnitude of P^* that is the magnitude of the longitudinal compressive force shortly after the shock section:

$$\frac{P^*}{\sigma\omega^2 R^2} = \frac{1}{1 + \varepsilon^*}\left(1 - \frac{1}{\omega}\frac{d\theta^*}{dt} \right)^2 - \left(1 + s^* - \frac{1}{\omega}\frac{d\theta^*}{dt} \right)^2$$

and, by following the equation 8, since $\left| \dfrac{1}{\omega} \dfrac{d\theta *}{dt} \right| << 1$, $\left| s * \right| << 1$ and $\left| \varepsilon * \right| << 1$, it follows:

$$\frac{P*}{\sigma \omega^2 R^2} = \varepsilon *.$$ (5.9)

The previous relation (equation 5.9) shows a direct proportionality between the compression force $P*$ and $\varepsilon *$, therefore, since the kinematical strain $\varepsilon *$ is always very little, the compression force $P*$ (shortly after the shock section) is negligible, namely on the idle arc $\kappa_{idle} \cong 1$.

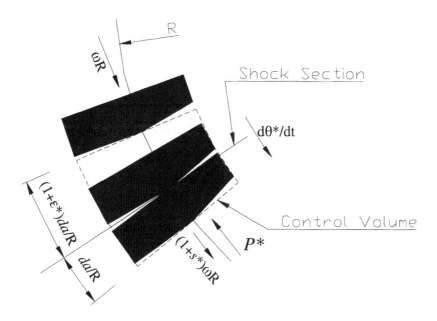

Fig. 5.6 Shock section dynamics

5.8 Integration of equations

Two different situations have to be considered: driver pulley and driven one. On the driver pulley the metal plates are separated on the idle arc, while they are pressed together on the active arc whose extension goes from the shock section to the exit point of the belt. Therefore in this case it is convenient to integrate the congruency and the first momentum equations (equations (1)–(2)), on the active arc, starting from the shock section. While the starting point of integration, on the driven pulley, will be the real or imaginary point where the sliding angular velocity vanishes.

It is useful to change the angular co-ordinate by means the following transformation rule $\psi = \theta - \theta *$, the congruency equation (equation 5.1) gives:

$$s = s* - w\psi \tag{5.10}$$

where $s*$ is the sliding coefficient shortly after the shock section ($\theta = \theta*$) on driver pulley and it is zero on the driven pulley where $\theta*$ is defined as the point with null sliding coefficient. The previous equation together with kinematic equation (equation 4) allows to write the subsequent relation:

$$\tan\gamma = \tan\gamma* - \psi \tag{5.11}$$

Where $\gamma*$ is the sliding angle for $\theta = \theta*$. In order to integrate the first momentum equation it is convenient to introduce the new variable $\delta = \psi - \tan\gamma*$.

Observe that $\cos\beta_s$ is always positive so, by taking into account the geometric equation (equation 5.5), it is possible to write:

$$\cos\beta_s = \frac{1}{\sqrt{1+\tan^2\beta_s}} = \frac{1}{\sqrt{1+\tan^2\beta\cos^2\gamma}}$$

Two cases have to be distinguished: $w>0$ (increment of pitch radius) and $w<0$ (decrement of pitch radius).

First Case: $w>0$

When the pitch radius of the belt grows the sliding angle γ belongs to the interval $\left]-\frac{\pi}{2},0\right]$ so that $\cos\gamma$ is always positive and it is possible to write $\cos\gamma = \frac{1}{\sqrt{1+\tan^2\gamma}}$. With this position the first momentum equation (equation 5.2) becomes: $\dfrac{\partial\kappa}{\kappa\partial\delta} = \dfrac{\delta}{1 - \dfrac{\tan\beta}{\mu}\sqrt{1+\delta^2\cos^2\beta}}$

The previous equation, once placed $\kappa* = 1$, can be integrated analytically to give:

$$\kappa = \frac{G_1(\psi - \tan\gamma*)}{G_1(-\tan\gamma*)} \tag{5.12}$$

where $G_1(\delta) = e^{H_1(\delta)}$ and

$$H_1(\delta) = -\frac{\mu^2}{\sin^2\beta}\left(\frac{\tan\beta}{\mu}\sqrt{1+\delta^2\cos^2\beta} + \ln\left|1 - \frac{\tan\beta}{\mu}\sqrt{1+\delta^2\cos^2\beta}\right|\right)$$

Second Case: $w<0$

When the pitch radius of the belt diminishes the sliding angle γ belongs to the interval $\left[\pi,\frac{3\pi}{2}\right[$ so that $\cos\gamma$ is always negative and it is possible to write $\cos\gamma = -\frac{1}{\sqrt{1+\tan^2\gamma}}$.

In this case the momentum equation (equation 5.2) becomes:

$$\frac{\partial \kappa}{\kappa \partial \delta} = \frac{\delta}{1 + \dfrac{\tan \beta}{\mu} \sqrt{1 + \delta^2 \cos^2 \beta}}$$

and the solution of the previous equation, once placed $\kappa^* = 1$, is:

$$\kappa = \frac{G_2\left(\psi - \tan\gamma *\right)}{G_2\left(-\tan\gamma *\right)} \tag{5.13}$$

where $G_2(\delta) = e^{H_2(\delta)}$ and

$$H_2(\delta) = -\frac{\mu^2}{\sin^2 \beta}\left(\frac{\tan \beta}{\mu} \sqrt{1 + \delta^2 \cos^2 \beta} - \ln\left|1 + \frac{\tan \beta}{\mu} \sqrt{1 + \delta^2 \cos^2 \beta}\right| \right)$$

5.9 Consideration on the solution of equations

5.9.1 Driver pulley

On the driver pulley the solution of equations is defined for positive values of angular co-ordinate ψ i.e. on the active arc, while on the idle arc, under the hypothesis of uniform kinematical strain, the net tension of the belt is constant together with the linear pressure. Now from the relation $s = s* - w\psi$ (equation 5.10) it follows that, when $w > 0$, the sliding coefficient decreases with ψ i.e. the sliding coefficient is always negative on the active arc.

For $w < 0$ the sliding coefficient grows with angular co-ordinate ψ so it is possible to have null sliding coefficient on the active arc, this happens for:

$$\psi = \hat{\psi} = \frac{s*}{w} = \tan\gamma * \tag{5.14}$$

Now we observe that equations 5.12 and 5.13 are symmetric with respect to $\delta = 0$ i.e. $\psi = \hat{\psi}$, therefore, when $w < 0$, the extension α of active arc cannot exceed the value $2\hat{\psi}$ in order to assure the pulley being driver: the dimensionless tension κ_2 in the exit point of the belt has to be lower than $\kappa_{idle} \cong 1$. Figure 5.7 represents equation 5.11 in the $[\psi, \tan\gamma]$ plane. It is evident that, when $w > 0$, $\tan\gamma$ is always negative and that, when $w < 0$, $\tan\gamma$ vanishes for $\psi = \hat{\psi}$ where the sliding coefficient goes to zero. Therefore we distinguish two region one for $w > 0$ and the other for $w < 0$. Figures 5.8 and 5.9 show the dimensionless tension versus angular co-ordinate ψ for different values of sliding angle $\gamma*$ and for the two cases: $w > 0$ and $w < 0$. During pitch radius increment the dimensionless tension is monotone that is it always decreases with ψ. Different is the case with pitch radius decrement: κ decreases first, until the point $\hat{\psi}$ (equation 5.14) is reached, then it grows up to unit value for $\psi = 2\hat{\psi}$. This phenomenon sets a restriction on the maximum value of dimensionless parameter w, in effect $\hat{\psi}$ grows with $\tan\gamma*$ so we can assert that, since the sliding coefficient $s*$ is always little, the non-dimensional radial velocity's maximum absolute value has to be little in order to assure high values of $tan\gamma^*$ and make the pulley driving especially for high transmitted torque. If the absolute value of w is not sufficiently little, the previous equation tells us that the pulley becomes driven and that the set of equations, necessary to describe the dynamics of the belt, must be changed. Conversely, when the pitch radius of the belt increases, no problem arises. Figure 5.10 shows the values of dimensionless tension κ_2 in the exit point of the belt for different values of the extension α of the active arc and for $w<0$. The figure underlines that as α grows the minimum value of γ^*, needed to assure κ_2 to be lower than $\kappa_{idle} \cong 1$ i.e. to guarantee the pulley being driver, grows.

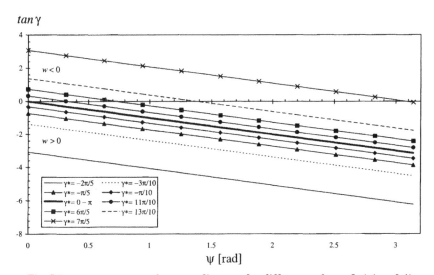

Fig. 5.7 $tan\gamma$ **versus angular co-ordinate ψ for different values of γ^* ($\mu = 0.1$)**

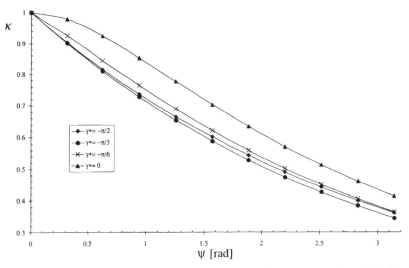

Fig. 5.8 **dimensionless tension κ versus ψ for different values of γ^* ($w>0$, $\mu = 0.1$, $\beta = 18°$)**

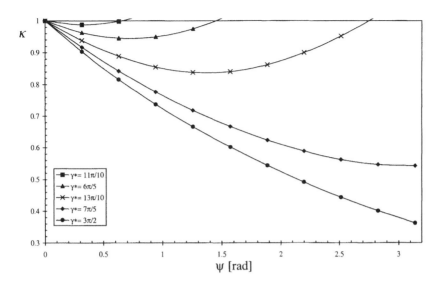

Fig. 5.9 **Dimensionless tension κ versus ψ for different values of γ^* ($w<0$, $\mu = 0.1$, $\beta = 18°$)**

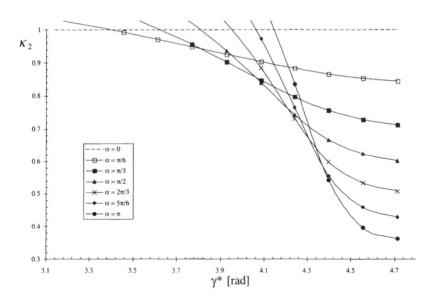

Fig. 5.10 κ_2 **versus γ^* for different values of the active arc's extension α ($w<0$, $\mu = 0.1$, $\beta = 18°$)**

5.9.2 Driven pulley

The metal plates are pressed together on the driven pulley where no clearance between plates exists. Therefore the dynamical behaviour of the belt depends only by the position of the point with null sliding tangential speed. Figure 5.11 represents the relative motion of the belt in respect to pulley in transient condition with pitch radius decrements. There are visible two belt's positions: the first at time t and the second at $t+dt$. Moreover, the picture shows the relative displacements of the belt and the relative velocity with its radial and tangential components, respectively $\dot{R}$ and $\omega_s R$. Let observe that the radial component of the sliding velocity does not depend on the angular co-ordinate, while the sliding angular velocity varies linearly with the ψ angle. The figure shows the arc HF being longer than HM arc, consequently the two relative displacement, along the ξ direction, of points F and M differs, giving a different from zero mean sliding coefficient.

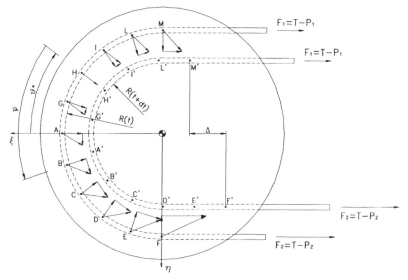

Fig. 5.11 Metal V-belt motion (driven pulley $w<0$)

The trend of sliding angle γ is depicted (Fig. 5.12) for $w>0$ (pitch radius increment) and $w<0$ (pitch radius decrement).

Figure 5.13 shows the graph of dimensionless tension versus angular co-ordinate ψ for $w>0$ and $w<0$, the two lines are symmetric with respect to the point $\theta*$ i.e. $\psi=0$. The trend of dimensionless tension κ greatly differs for the two cases. In effect when the pitch radius increases the dimensionless tension assumes the maximum value for $\psi=0$, conversely when $w<0$ no maximum exists and the dimensionless tension κ assumes the minimum value in the same point $\psi=0$ namely $\theta=\theta*$.

It must be highlighted that, for a fixed value of sliding angle γ or angular co-ordinate ψ, the dimensionless tension κ does not depend on the value of dimensionless radial velocity w, but on the sign of this quantity, i.e. on the sign of $\dot{R}$. The dimensionless radial velocity affects only the value of local sliding coefficient s and influences the value of $\theta*$.

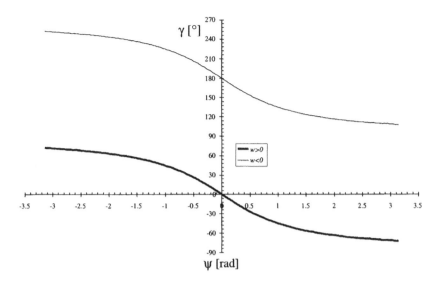

Fig. 5.12 Sliding angle γ versus angular co-ordinate ψ (driven pulley)

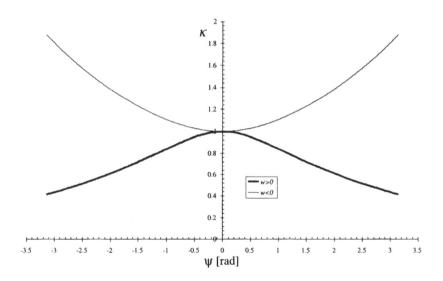

Fig. 5.13 Dimensionless tension κ versus angular co-ordinate ψ
(driven pulley, μ = 0.1, β = 18°)

Now let us observe again Fig. 5.11, where the belt's tensions of points F and M appears to be different. In effect we have, previously, said that the dimensionless tension κ is symmetric in respect to co-ordinate ψ = 0, namely θ = θ *, hence, owning to the greater length of HF arc in respect to HM arc, the following result is achieved: $\kappa(\mathrm{F}) > \kappa(\mathrm{M})$ (Fig. 5.6), namely $F_2 > F_1$, so that the transmission of torque is allowed.

5.10 Conclusions

In this work we propose a theoretical model of transient dynamics of metal pushing V-belt with clearance between plates. On the driven pulley this kind of belt behaves like the belt without clearance, while on the driver pulley the dynamical behaviour differs greatly: there exist a shock section, where the plates bring into collision, which subdivide the contact arc into two part: an idle arc (from entry point to shock section) and an active arc (from the shock section to the exit point of the pulley). On the idle arc no tangential slip exists between pulley and blocks and the plates are separated: due to absence of tangential slip no torque is transmitted on this arc. On the active arc the clearance is null, longitudinal compression forces and tangential slip exist in order to transmit torque.

Different mechanical behaviour of the belt arises respectively for pitch radius increment and decrement. In the first case, on the active arc, the dimensionless sliding coefficient, i.e. the tangential slip, is always negative and diminishes together with dimensionless tension as the angular co-ordinate grows: no problem arises. In the second case, with pitch radius decrement, one point with null sliding coefficient could exist on the active arc, consequently the dimensionless tension first would reduce until its minimum value is reached and then would increase. Since the pulley is driver i.e. the dimensionless tension in the exit point is lower than that in the shock section, the previous phenomenon implies the active arc to be lower than $\dfrac{2s*}{w}$ that is the dimensionless radial velocity w is constrained to assume little values.

Further development of the work will be made in the next future by means experimental investigation, which will be necessary to verify the proposed model. Moreover, the effect of pulley and belt strain motion will be consider in order to examine its influence on shifting dynamics of the belt.

5.11 References

(1) **Paul, M.,** (1999) CVTs driving the future of transmission technology. Int. Congress on *Continuously Variable Power Transmission CVT '99*, 16–17 September, 1–8. Eindhoven, The Netherlands.

(2) **Mangialardi, L.** and **Mantriota, G.,** (1992) The advantages of using continuously variable transmission in wind power systems. *Renewable Energy,* **2** (3) 201–209.

(3) **Mangialardi, L.** and **Mantriota G.,** (1992) Continuously variable transmission with torque-sensing regulators in waterpumping windmills. *Renewable Energy,* **7** (2) 807–823.

(4) **Mangialardi, L.** and **Mantriota G.,** (1994) Automatically regulated C.V.T. in wind power systems. *Renewable Energy* **4** (3), 299–310.

(5) **Mangialardi, L.** and **Mantriota G.,** (1999) Power flows and efficiency in infinitely variable transmissions. *Mechanism and Machine Theory,* **34**, 973–994.

(6)   **Mangialardi, L.** and **Mantriota G.,** (1996) Dynamic behaviour of wind power systems equipped with automatically regulated continuously variable transmission. *Renewable Energy,* **7** (2) 185–203.

(7)   **Arnijima, S., Fujii, T., Matsuoka, H.,** and **Ikeda, E.,** (1992) Study on axial force and its distribution of a new CVT Belt for car. *Int. Journal of Vehicle Design,* **13** (2) 168–181. Interscience Enterprises Ltd.

(8)   **Gerbert, B. G.,** (1984) Metal V-Belt Mechanics. *ASME Journal of Mechanical Engineering,* 84-DET-227.

(9)   **Kim, H. and Lee J.,** (1994) Analysis of belt behavior and slip characteristics for metal V-Belt CVT. *Mech. Mach. Theory,* Elsevier Science, **29** (6) 865–874.

(10) **Sun, D. C.,** (1988), Performance analysis of variable speed-ratio metal V-Belt drive. Transaction of the ASME, *Journal of Mechanism, Transmission, and Automation in Design,* **110,** 472–481.

(11) **Ide, T., Udagawa, A.,** and **Kataoka, R.,** (1995) Simulation approach to the effect of the ratio changing speed of a metal V-Belt CVT on the vehicle response. *Vehicle System Dynamics,* **24** 377–388.

(12) **Kanehara, S., Fujii, T.,** and **Oono S.,** A study on a metal pushing V-Belt type CVT (macroscopic consideration for coefficient of friction between belt and pulley), *JSAE CVT '96 Yokohama Proceedings,* 15–22.

(13) **Kanehara, S., Fujii, T.,** and **Fujiimura, O.,** Characterization of CVT using a metal V-Belt at transitional states. Int. Congress on *Continuously Variable Power Transmission CVT '99,* 16–17 September, 58–64, Eindhoven, The Netherlands.

(14) **Carbone, G., Mangialardi, L.,** and **Mantriota G.,** (2001) Theoretical model of metal V-Belt drives during rapid ratio changing. *ASME Journal of Mechanical Design,* Vol. **123**.

(15) **Carbone, G., Mangialardi, L.,** and **Mantriota, G.,** (2000) "Influence of clearance between plates in metal pushing V-Belt dynamics", sent to ASME *Journal of Mechanical Design.*

6

Cylinder Balancing Control of Direct Injection Engines

G N Heslop and **J Dixon**

Abstract

This chapter presents a method of balancing the cylinder-to-cylinder torque fluctuation of direct injection (DI) engines by controlling the individual cylinder fuel quantity. Known as cylinder balancing, this algorithm has been implemented on drive-by-wire four-cylinder diesel vehicles and three-cylinder DI gasoline vehicles. It is shown to improve the engine idle speed regulation, to have the potential to reduce the passenger compartment noise and vibration, and to provide some diagnostic capability.

The approach is to estimate the accelerations produced by firing engine cylinders based upon a non-contacting crankshaft position measurement. These measurements are used to learn appropriate cylinder offsets to correct this fluctuation. Offset learning times achieved are typically a few seconds. Experimental verification of the method is presented.

6.1 Introduction

Cylinder balancing is a method for equalizing the torque output of each cylinder. Without it, cylinder imbalance may be detectable by the human ear at low engine speed and can result in vibration which can be felt by occupants in the vehicle passenger compartment. If one cylinder produces less torque than the other three on a four-cylinder engine, the resulting engine speed frequency spectrum will exhibit a peak magnitude around ¼ firing frequency. This frequency is about 7 hz in idle operation, a low frequency which may excite some components in the engine or vehicle, resulting in undesirable noise and vibration (NVH).

In steady engine operating conditions, the chief cause of cylinder torque imbalance is believed to be the different injector characteristics. There is, of course, no direct feedback about the actual fuel quantity delivered by each injector. Differences in the combustion characteristics of the individual cylinders resulting from an inequitable intake of air and recycled exhaust

(EGR) charge, different cylinder operating temperatures, different intake runner lengths, soot deposit build up on the valves, and so on, also contribute to these torque differences.

6.2 Cylinder balancing for diesels

The cylinder balancing algorithm was initially designed for implementation on a DI four-cylinder drive-by-wire diesel engine, fitted with a distributor fuel pump. The engine management controller (EEC) executes the algorithm, using as its prime input crank position data, and outputs a fuel mass demand request over CAN to a pump controller (PCU), which implements the request. The fuel quantity injected can be controlled on a cylinder-by-cylinder basis over the full range of speed operation of the engine.

The algorithm has been implemented on a number of systems of this type. With some minor changes it has also been applied to diesel common rail systems and a DI gasoline vehicle.

Figure 6.1 gives a simplified context diagram showing where the cylinder balancing algorithm resides. The main fuel controller arbitrates between the idle- and pedal-based fuel controllers, producing a main fuel mass demand fqd_fuel_dmd. The cylinder balancing feature produces a cylinder resolved fuel mass offset. The sum of these two is sent in a CAN message to the PCU, at a crank angle which is dependant on engine speed. Messages have to be sent a greater number of crank degrees before the next firing event as the engine speed rises, to ensure the pump controller can implement the request at the next injection.

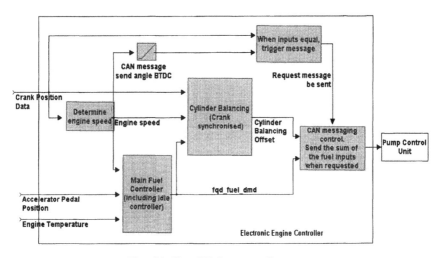

Fig. 6.1 Simplified context diagram

6.2.1   Selection of crank position signal

The approach is to estimate the accelerations produced by the firing engine cylinders, based upon a non-contacting crankshaft position measurement. This sensor typically provides position information every 10 degrees crank or finer, depending on the granularity of the fitted toothed wheel. A 36-2 toothed wheel has been fitted to the engines for which results are presented in this paper. Using crank position data to provide individual cylinder operating behaviour was the preferred approach, as it did not involve additional costs. Other approaches using cylinder pressure sensors or torque sensors may be possible in future.

6.2.2   Selection of crank window

Figure 6.2 shows a recording of engine speed data from a diesel engine sited within a dynamometer facility, fitted with an optical encoder providing engine position every degree. Similar data was also recorded on an engine with one out-of-specification injector fitted, having a different injector nozzle opening pressure. The crank position data and cam signal indicating which firing event corresponded to cylinder 1, were used in simulations to design a crank position data processing scheme and to verify that it identified the cylinder fitted with the out-of-specification injector.

The injection timing could vary between 20 degrees before top dead centre (BTDC) and 20 degrees after top dead centre (ATDC). The data processing algorithm avoids using data pulled from this region, as it is subject to large transients. As an acceleration estimate was required for each firing event, the speed before and after each firing event, based on the rotation period from 60 degrees BTDC to 20 degrees BTDC, was selected and divided by the 180 degree rotation period (see Figs 6.2 and 6.3). This crank region is a fairly steady condition in the engine cycle, considered a good choice, and verified by simulation. The speed calculated after a firing event is used as the speed prior to the next firing event, so only one speed is required to be calculated every 180 degrees of crank rotation.

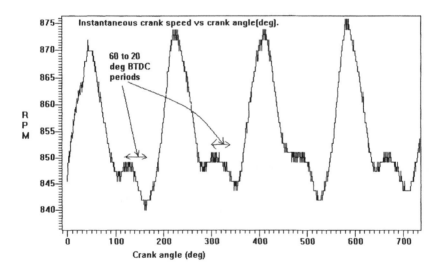

Fig. 6.2 Recording of instantaneous engine speed

6.2.3   When to learn/apply/forget the offsets

The estimated accelerations are used to provide correcting fuel mass offsets which are added to the main fuel quantity demand on an event by event demand basis. Learning of the fuel mass offsets only occurs when a number of conditions simultaneously exist.

One of these conditions is that the fuel (torque) demand must be considered steady, when it may be expected that the accelerations observed following firing events would be equal. In addition, learning of the offsets only occurs below a calibrated speed value, when the benefits of cylinder balancing are likely to be appreciated. This is because any cylinder imbalance, results in an engine-speed, spectral density peak occurring at a frequency which is proportional to engine speed. Higher spectral frequencies tend to result in less NVH. If any offset, when added to the main fuel quantity, would result in a negative fuel demand, learning and implementation of the desired offsets is terminated. The offsets are, however, held in memory and learning and implementation of them resumes when more suitable operating conditions occur.

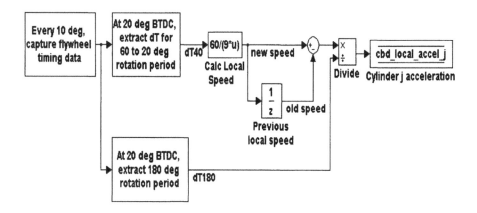

Fig. 6.3 Individual cylinder acceleration estimate

When learning is occurring, the offsets are always being implemented, to avoid having an unstable system. A response is expected to any adjustment of the offsets. If this fails to materialize, the offsets will wind up or down. If learning ceases for a period of time, the offsets are eventually reset to zero, as they may no longer be suitable corrections.

6.2.4   Cylinder balancing authority
The authority given to cylinder balancing is limited, typically to a few mg/stroke, to prevent unlimited wind-up of the offsets in the condition that acceleration offsets persist between the reference acceleration and one or more cylinder event accelerations. This could occur if, for instance, an injector became blocked.

6.2.5   Calculation of the offsets
The EEC has no knowledge as to which cylinder in the block is to fire next, as the cam information available to the PCU was not shared. This information was not required to implement the algorithm as the firing order is fixed at 1342 when the cylinders are numbered 1234 in the block. The algorithm selects one cylinder as the reference cylinder to which it attempts to make the accelerations of the other three cylinders equal to that of the reference cylinder (in learning conditions). The reference cylinder is therefore assumed to be operating correctly and offset corrections are learned for the other three cylinders. Each cylinder balancing offset is updated once every engine cycle (every four events) when in a learning state. If the reference cylinder is referred to as 0 and the firing order as 0123, the cylinder accelerations are updated after each firing event at 20 degrees BTDC. Given a sequence of

firings of 01230123, the cylinder balancing offset for cylinder 3 is updated after the reference cylinder 0 has fired as these events are consecutive. The offset for cylinder 1 is updated after cylinder 1 has fired and the offset for cylinder 2 after cylinder 2 has fired. Integral controllers exist to update each offset as shown in Fig. 6.4. A higher integral gain reduces the time to learn, but it hasn't to be chosen so large as to interfere with the idle control.

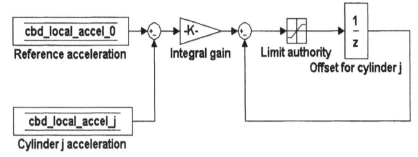

Fig. 6.4 Update of fuel mass offset for cylinder j

6.2.6 Measured data – comparison with and without cylinder balancing

Cylinder balancing performance data was recorded in a Mondeo vehicle installed with one of a family of new diesel engines. The engine speed, the idle fuel mass demand, and the cylinder balancing offsets were recorded every 20 ms in an idle operating condition, with and without cylinder balancing in operation. Cylinder balancing was disabled by setting the offset limit authority (shown in Fig. 6.4) to zero.

The first row of Fig. 6.5 shows the results with cylinder balancing off and below with cylinder balancing on. The engine speed variation is reduced with cylinder balancing in operation. The spectral content of the engine speed signal was analysed in each case. With cylinder balancing disabled, there are peaks in the power density at ¼ and ½ firing frequencies (7.5 hz and 15hz at 900 r/min). When cylinder balancing was enabled, these peaks were practically eliminated. The offsets learned when cylinder balancing was enabled, indicate that two cylinders were behaving similarly, producing more torque than the chosen reference and one cylinder was down in power. As the cylinder balancing fuel offsets average to a negative fuel mass, the operation of cylinder balancing had the effect of requiring the idle controller to demand a higher fuel quantity when cylinder balancing was on, to maintain the idle speed. With cylinder balancing on, the idle fuel mass demand appears to be controlling with less effort, as a result of the cylinder torque differences being compensated for by a separate controller.

The qualitative improvement in NVH with cylinder balancing in operation is a subtle effect.

6.2.7 Measured data – offset learning time

Figure 6.6 shows the effect of integral gain value on the offset learning times, at an idle operating condition. The figure illustrates that a higher gain reduces the learning time.

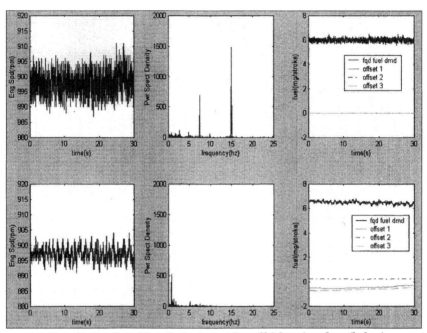

Fig. 6.5 Results with cylinder balancing off (above) and on (below)

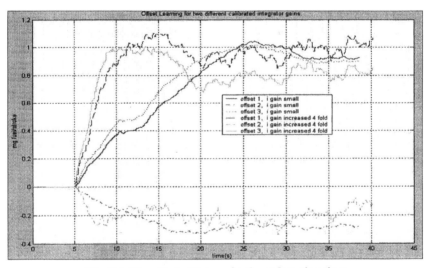

Fig. 6.6 The effect of integral gain on learning time

6.2.8   Measured data – when to learn/apply/forget the offsets

Figure 6.7 shows the effect of the driver 'blipping' the accelerator pedal on the cylinder balancing algorithm. In the case shown, when the transient conditions occur, learning is terminated, and a timer starts incremented to record the period of time for which learning has ceased. In addition, when a certain engine speed is exceeded (approximately 1200 r/min), the offsets are no longer applied, but are still saved in memory. On return to a lower engine speed, learning of the offsets resumes from the previously saved offsets, as the timer did not reach the value at which the offsets would be forgotten.

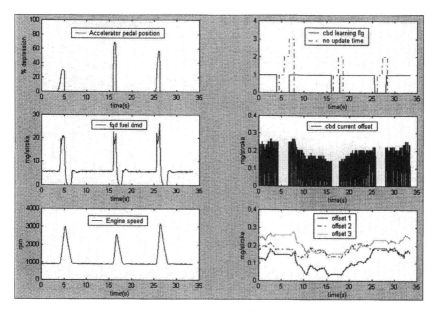

Fig. 6.7 Effect on cylinder balancing of 'blipping' the accelerator pedal from an idle condition

6.2.9   Design limitations

At each engine key-on, a different cylinder in the block may be chosen as the reference cylinder, as without a cam signal, the EEC has no knowledge as to which cylinder in the block is to fire next. A consequence of this is that the offsets learned cannot be stored in memory for use at subsequent key-on's. The offsets are learned from scratch at each engine start-up, which can take a few seconds depending on the integral gain.

6.3 Cylinder balancing for DI gasolene engines

The cylinder balancing algorithm was modified for implementation on a DI three-cylinder drive-by-wire gasoline vehicle installed in a Fiesta. The control system for this demonstrator vehicle was implemented using CACSD (Matlab/Simulink) and rapid prototyping tools (dSPACE), whereas the diesel implementation was within a production EEC.

6.3.1   Choice of actuator

The engine operates in two combustion modes, stratified and homogeneous modes. The main strategy controlling the engine is torque based, chiefly because combustion mode changes have to be transparent to the driver which requires accurate control of the torque. The torque is a function of about five variables and of these variables, only fuel mass and spark timing were controlled on an event by event basis.

In stratified mode, the spark timing has little effect on the torque, so fuel mass was chosen as the primary actuator for the main torque control and implementation of cylinder balancing.

In homogenous mode, it is important to maintain a particular AFR, so the main fuel mass demand is a slave to the intake mass of air. There was also a closed-loop AFR controller adjusting the overall fuel mass. It was decided that in homogeneous mode, cylinder balancing would initially try balancing with fuel mass offsets. If a fuel offset reached the prescribed limiting value, balancing with spark timing would be chosen instead. However, spark timing adjustments were not exclusively reserved for cylinder balancing, as spark adjustments are required for transient torque control, for example at idle.

During mode changes, when the air flow, fuel, and spark are changing to achieve the torque requirements, cylinder balancing is turned off.

6.3.2   Control system architecture

Figure 6.8 shows a simplified sketch of the control system architecture. The cylinder balancing feature bypasses the torque control system and the spark and fuel offsets are added directly to the main fuel and spark timing requirements.

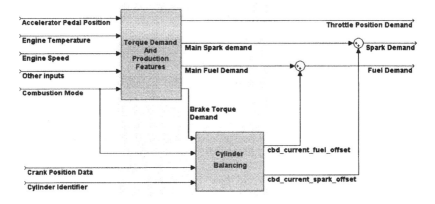

Fig. 6.8 Simplified sketch of the control system architecture

6.3.3   Differences from diesel

On a three-cylinder engine, combustion events occur every 240 degrees. An internal company report suggested a few changes to the diesel algorithm, some of which were adopted on this application, and recounted below.

One of the suggestions was to use the crank window 100 to 170 degrees ATDC of each cylinder to determine the speed after each event. This data was used in a similar manner on diesel, to determine the individual cylinder acceleration estimates.

Another change adopted was to use a reference acceleration which is the average of the three-cylinder accelerations. The benefit of this is that the offsets learned are not so sensitive to one poorly performing cylinder. The acceleration errors were integrated as before, to form the offsets. This algorithm therefore has an offset for every cylinder, one more than the algorithm applied on diesel. The integrator gain used was made dependant on the combustion mode, and actuator choice because the torque sensitivities very with these.

One further change adopted was to force the average cylinder offset to be zero, by subtracting off the average cylinder offset at each update. This was thought to reduce the likelihood of a torque step when the cylinder balancing offsets are reset to zero, which commonly occurs on mode changes. The authority given to cylinder balancing was again limited, the limit depending on the combustion mode and the actuator choice.

It was decided that if the given authority limit was breached, cylinder balancing would be abandoned with that particular actuator, as it had clearly failed to balance the engine within the constraints set. Balancing with the other actuator was then enabled, and if that failed in a similar manner, cylinder balancing was abandoned and is only restarted following a mode change or an ignition key-off/key-on sequence occurring.

The offset update algorithm is summarized in Fig. 6.9. Although not used, there is the facility to ensure the average offset equals a calibrated bias value. This might be useful, for example, in homogenous mode when the spark timing might otherwise operate at a point of maximum advance, but to implement cylinder balancing, freedom to advance and retard spark is required. Setting a bias of a few degrees spark retard would enable cylinder balancing to operate, but the benefits of doing this would need to be traded off against a reduction in fuel economy.

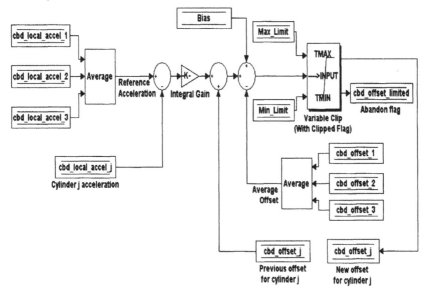

Fig. 6.9 Update of the offset for cylinder j

The decision as to when cylinder balancing should learn is based on whether the brake torque is steady, as opposed to just the fuelling level on diesel.

6.3.4   Measured data – homogeneous mode

The development of the torque-based engine idle controller consumed the majority of development time in the homogeneous combustion mode. Consequently, there are no results to present on cylinder balancing in this mode. It is fair to say that cylinder balancing in this mode is a greater challenge than in stratified mode, because fuel and spark are so busy meeting the demands of other control systems.

6.3.5   Measured data – stratified mode

Stratified idle data with and without cylinder balancing in operation was recorded (see Fig. 6.10). The engine speed spectral content contained peaks at 1/3 and 2/3 firing frequencies when it was turned off. These are consistent with the offsets which were learned to cause their removal. The engine speed signal was cleaned up as a result of cylinder balancing executing. There is no net change in the main fuel demand with and without cylinder balancing, which is to be expected if the cylinder balancing offsets sum to zero.

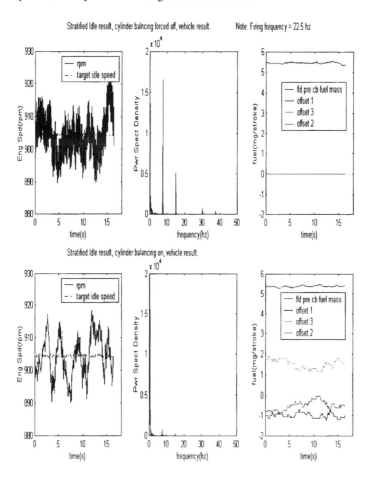

Fig. 6.10 Comparison of stratified idle data with cylinder balancing off (above) and on (below)

6.4 Conclusions

Cylinder balancing has been demonstrated to successfully operate on two different engine types of differing numbers of cylinders. On DI gasoline, the demonstration has been limited to stratified mode. Further development is required in homogeneous mode.

Very little calibration effort was required in those cases demonstrated, the chief parameter being the integrator gain, which governs the offset learning time.

If an offset grows to a particular size, this information could be used to diagnose an engine malfunction.

7

Continuously Variable Transmission with Electromechanical Power Splitting

G Avery and **P Tenberge**

7.1 Introduction

A modern transmission should have the following qualities.

- A large range of ratios to allow for good acceleration as well as good fuel consumption.
- Continuously variable operation, providing smooth operation and optimized use of the motor's efficiency and performance potential.

Furthermore, some tasks usually reserved for other parts of the driveline can be integrated, such as:

- regenerative braking;
- starter and alternator function.

Before developing a new transmission, it is useful to take a look at the state of the art.

7.1.1 Automatic transmissions
The newest developments on this front are the six-speed automatic transmissions, for example, the ZF transmission from ZF based on the LePelletier design. In comparison with previous automatic transmissions, this transmission has a higher range of ratios (approx. 6:1), increased efficiency and shifts at least as comfortably as other automatic transmissions, which is to say almost imperceptibly in some cases.

7.1.2 Manual transmissions
A demand for greater efficiency and performance has led to the six-speed manual transmission. In the proper hands these transmissions can indeed improve fuel consumption as well as driving enjoyment, but many drivers simply do not know what to do with so many gears. Therefore, the shifting is left to what essentially is a robot, an electronic processor with pneumatic and/or hydraulic actuators. Shift comfort is perhaps not as good as with automatic transmissions, but efficiency is higher in the absence of clutch packs and a torque converter.

7.1.3   Mechanical continuously variable transmissions

By virtue of their ability to run the engine at almost any operating point regardless of vehicle speed, these transmissions can milk the absolute last reserves of power and efficiency out of an engine. Currently, the two most popular types are belt/chain-drive CVTs as in the Audi A6 Multitronic and toroidal traction-drive designs as in the Nissan Gloria/Cedric (1) (only available in Japan).

Although CVTs allow the selection of the optimum motor operating point, the higher losses in the transmission itself compared to other types of transmissions, partially counteract the gains from the better use of the motor. In some cases, losses at higher vehicle speeds can cause cooling problems which necessitate an artificially low top speed. Another stumbling block is that CVTs, while admittedly in their infancy, are considerably heavier than their conventional counterparts.

A further problem is that the motor makes sounds unaccustomed to the driver. Although most people quickly get used to the smooth rises and dips in engine speed, many manufacturers provide 'virtual gears' by programming the CVT to operate at discreet ratios like a conventional transmission.

7.1.4   Electro-mechanical continuously variable transmissions

A common characteristic of these transmissions is that part or all of the power is converted to electricity and then back into mechanical power. This immediately offers at least two attractive possibilities: first, the elimination of the alternator, and second, easy storage of this energy, for example in conjunction with regenerative braking. Furthermore, since powerful electric motors are in place anyway, it is a fairly simple matter to use one of these motors to start the prime mover, the internal combustion engine.

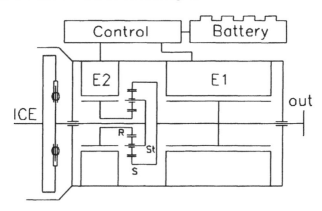

Fig. 7.1 Toyota Hybrid System

The Toyota Hybrid System (THS) installed in the Toyota Prius (available now in Japan and as of September 2000 in Europe) is perhaps the most well-known and simplest example of such a system (2), (3), (4). A large electric motor E1 is situated directly on the output shaft and is connected to a smaller electric motor, E2, via a power control module. By sending the majority of the power through the epicyclic gear (superimposition transmission) the losses in the electric branch are reduced, and the limitless (+/- ∞) range of ratios in the electrical branch is brought down by mechanical limitations to a more usable final range of ratios.

The diagram below is an 'RPM ladder', with the geometry determined such that the points representing the rotational speeds of the transmission elements are always collinear. The torques of all elements are also related by a constant proportion just as if the arrows were forces acting upon a lever. To start the IC engine, E2 generates a positive torque which is braced by E1. Under normal driving conditions, E2 generates a negative torque by acting as a generator and the resulting electric energy is then sent through the power control module and converted back into mechanical energy by E1. Given a propeller shaft RPM and a target IC engine RPM, the rotational speed of E2 can be easily calculated and set, resulting in continuously variable operation. Finally, E1 can send braking energy to the battery as well as act as a booster for short spurts by drawing on this stored energy.

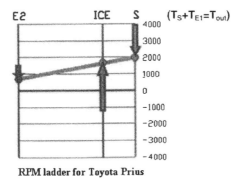

RPM ladder for Toyota Prius

Fig. 7.2 RPM ladder for Toyota Prius

The THS, while mechanically quite elegant, has the drawback that the final output torque can be no greater than about 70 per cent (depending on planetary gear ratio) of the available IC-motor torque plus the torque from E1. In order to increase final output torque to a 'sporty' level, IC or electric motor size must be increased to impracticable proportions, effectively limiting the system's use to small economy cars.

7.2 SEL-120 concept

An improvement upon the electrical power-splitting concept involves adding operating ranges, analogous to transmission speeds, to an electro-mechanical CVT in order to limit the amount of power flowing through the electrical branch, thereby improving efficiency and reducing the necessary size of the electric motors (**4**), (**5**). In the SEL-120, the generator E2 is connected to the sun gear as in the THS, but E1 is now connected directly to the IC engine and is the same size as E2.

The transmission as a whole is divided into two halves; the variator transmission and the superimposition transmission (**6**). Shaft speeds and torques produced or altered in the variator transmission are combined in the superimposition transmission to provide a final, continuously variable output.

Power is transmitted into the variator transmission through a dual mass flywheel, which decouples the driveline from the oscillations of the IC engine. Electric motor E1 is mounted directly on the input shaft from the IC engine, allowing it to also function as a starter, and the shaft is equipped with a one-way clutch used for driving forwards in pure electric mode. E2 is connected to a shaft coaxial with the main input shaft. Each electric motor is made of a rotor and water-cooled stator plus a position and speed sensor.

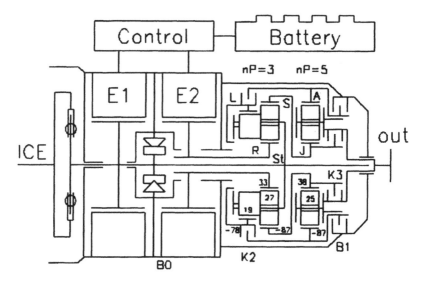

Fig. 7.3 Schematic drawing of the SEL 120/3

Behind the variator transmission is the superimposition transmission containing two epicyclic sets, the first a Ravigneaux four-shaft set and the second a simple epicyclic three-shaft set. Sun gear R, carrier St, fast ring gear S (almost always rotates faster than St) and slower ring gear L ($\omega L < \omega St$) make up the first stage. The second stage is comprised of sun gear J, carrier out, and ring gear A.

In operating range one, ring gear A is held stationary by brake B1, making the second stage a simple reduction gear. In operating range two, ring gear A is linked to ring gear L from the first stage, such that the first and second stages split the superimposition duties. In operating range three, the second stage is bypassed altogether by clutch K3, making ring gear S from the first stage the sole output.

To demonstrate exactly how the transmission works, it is useful to turn back to the RPM ladder, with the shaft speeds in each stage shown by the points along the thick lines and torques represented by arrows. Engaged clutches and brakes are marked by thick black lines around the appropriate points. The arrows along vertical St representing torque from the IC engine are lighter than the arrows for E1.

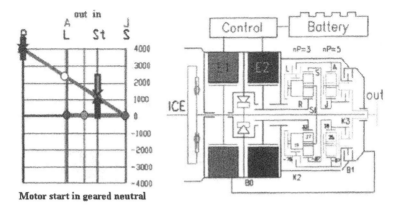

Fig. 7.4 Motor start and schematic

In this diagram, the IC engine is being started by electric motor E1. In order to remain still in geared neutral, motor E2 must also be accelerated to overcome its own inertia and that of associated shafts and gears.

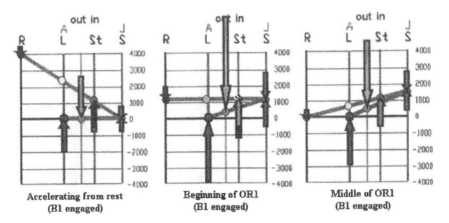

Fig. 7.5 Operating range 1

In the first diagram, the vehicle is being accelerated from rest. E2 is acting as a generator and the power coming from it re-enters as mechanical power through E1. As the vehicle is being accelerated in the next two diagrams, the speed of E2 is brought down so as to reduce the amount of power being transmitted electrically, thereby improving efficiency.

The shift from one operating range to another presents an interesting challenge because it requires a reversal in the torque coming from the electric motors, which must be carefully co-ordinated with the closing of the clutches. To maintain a torque equilibrium, the speeds are controlled such that shaft L is rotating about 50 RPM faster than shaft A (**7**). This speed difference causes clutch K2 to start producing a positive torque as its pressure is increased, relieving E2 while at the same time altering the torque equilibrium. This upset in the torque equilibrium results in a slight upset in the RPM balance, which the speed controller tries to compensate by changing the torque being delivered to both motors. By reacting to the change in torque-RPM balance caused by the closing of clutch K2, the torque delivered by the electric motors is altered such that the torques equilibrium is maintained.

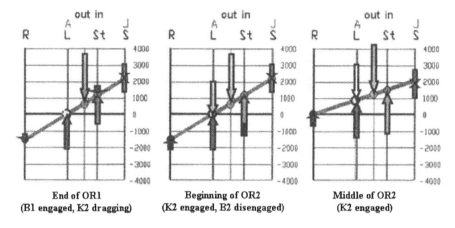

Fig. 7.6 Operating range 1–2

The torque being delivered by the electric motors is known from the amount of current running to them, so it is possible to calculate when brake B2 is no longer bearing a load. When this point is reached, B1 is quickly opened and K2 quickly closed, completing the shift.

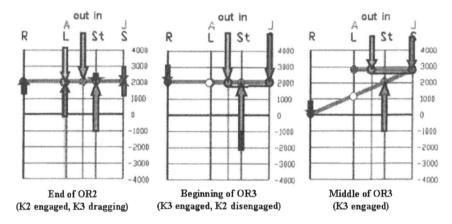

Fig. 7.7 Operating range 2–3

A similar trick is used in the shift from operating range 2–3. The electric motors are controlled such that sun gear J is rotating about 50 RPM faster than the output shaft. K3 is dragged to relieve E2, causing the regulator to react to the resulting upset in the RPM balance by adjusting the torques of the electric motors. It should be noted that a change in engine torque is not required to affect a shift.

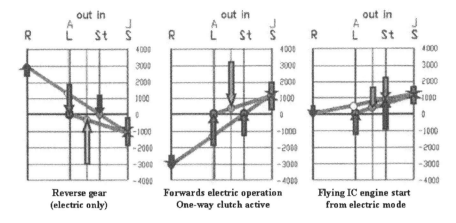

Fig. 7.8 Electric operation

Since electric motors, unlike IC engines, have no trouble running in reverse, there is no longer a need for an extra reverse gear. Torque is again created by using 'leverage' in the RPM ladder. Shaft A is held at one end by brake B1, so a torque in shaft S is multiplied. This torque in turn is provided by E2 acting against the main input shaft. Electric motor E1 can provide a bracing torque and if necessary, the drag torque of the engine can be used to assist. For forwards electric operation, this same principle is used, this time with a one-way clutch on the main input shaft. A fast transition from electric to IC operation is possible by starting the engine while moving. Here the available torque at the output shaft is somewhat lower than otherwise since E1 has to brace the driving torque while at the same time overcoming the drag torque of the engine.

7.3 SEL-120 design

Using the SEL-120 concept, a prototype-ready design was developed using the CAD software PRO/Engineer (6). Design work and simulations were carried out based on the following assumptions:

$T_{IC\ engine,\ max}$	= 300 Nm
$T_{E1,\ max}$	= 120 Nm
$T_{E2,\ max}$	= 120 Nm
$T_{output,\ max}$	> 1000 Nm

The electric motors used in this transmission are prototype, permanent, synchronous machines developed and built at the TU Chemnitz, each with a power rating of 10 kW long-term/25 kW sprint. The NiMH batteries have a capacity of 1.5 kWh at 300V, packaged in a suitcase-sized package weighing 46 kg. Only 60–70 per cent of this battery capacity is actually used, since charging and discharging in this range keep the battery's electrical efficiency above 90 per cent, and because available power drops off on the lower end of the charge level.

Battery capacity and the power rating of the electric motors were sized for good regenerative braking. Complete regeneration is as a practical matter impossible, since hard braking from a high speed sends several hundred horsepower through the brakes, a level which would require a vastly oversized electrical system to capture. Furthermore, regenerative braking without variable-brake pressure proportioning can result in a disturbance in vehicle stability if the electric motors exerted too high a torque on just one axle.

The final requirement was that the complete transmission fits in the transmission tunnel of a Mercedes-Benz C220 CDI. The concept lent itself well to being fit into a package very similar to modern automatic transmissions in dimensions and form.

SEL-120 in numbers

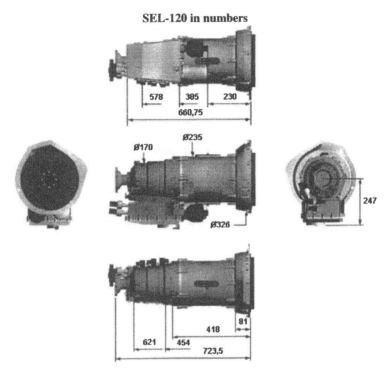

Fig. 7.9 Main dimensions of SEL-120

7.3.1   SEL-120 system weight compared to a similar CVT

Transmission weight with a 400 Nm IC motor

Electric motors (Rotors, Stators, Cooling)	47,3 kg
Superposition transmission + Housing	53,5 kg
Hydraulic control system	11,2 kg
Total	112 kg
Battery	46 kg
Power management electronics	20 kg
System weight (SEL)	**178 kg**
CVT of the same performance class	110 kg
Alternator	10 kg
Starter	8 kg
System weight (conventional)	**128 kg**

7.3.2   Range of ratios:

∞ (geared neutral) – 3.29 (operating range one) – 0.5 (maximum in operating range three)

For easy comparison, the schematic view of the transmission is shown below with a CAD-generated image of the complete design. The objects below the drive flange are sensors intended for the prototype only.

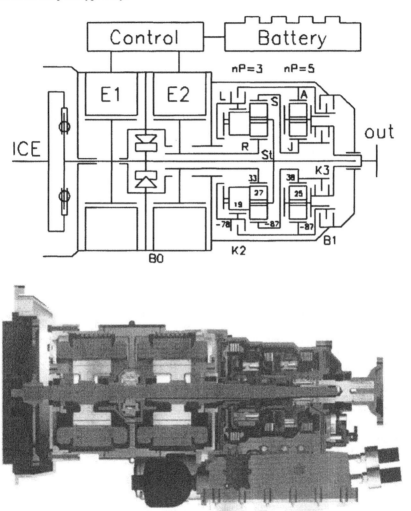

Fig. 7.10 and 7.11 Schematic and cutaway drawings of the SEL-120

Control system

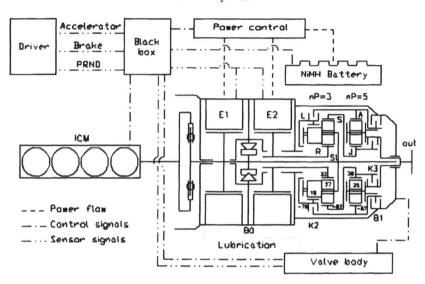

Fig. 7.12 Schematic drawing of SEL-120 control system

The hydraulic elements of the SEL-120 are similar to those of conventional automatic transmissions and do not require elaboration. New is the presence of electric motors and their co-ordination with the hydraulic system, all of which is directed by the 'black box', the microprocessor brain of the system. The wishes of the driver are interpreted based on signals coming from the accelerator, brake, and gear selector and translated into control signals to the transmission. This interpretation is not merely based on pedal position, but also pedal speed and a driver profile based on input information gathered throughout a certain time window. In this way, the control strategy is not only determined by the conventional sport/economy switch, but also by driver behaviour.

In ICE mode, transmission control follows IC engine control. A set of algorithms and maps relating the speed and acceleration of both engine and vehicle is stored in the microprocessor, and the proper algorithm is chosen according to the control strategy described above. In order to maintain a certain overall gear ratio, the electric motors are sent current in order to produce desired torque. If the resulting electric motor RPM deviates from the target RPM, the current is adjusted accordingly. This ability to automatically adjust torque in response to RPM deviation is used during the shift as described in section two.

The electric power control consists of DC/AC converters to convert AC electric motor voltage to/from the approximately 300 V DC battery voltage. Under cruise conditions, power flow is mostly between the two electric motors, with the battery serving mainly to store and reuse braking energy. Since the alternator is eliminated in vehicles with this system, it is also necessary for the electric power control to provide for the vehicle's electrical needs.

A simpler task for the control system is the operation in electric mode, in which only E2 is driven. This driving mode is especially useful for short distance (2–5 km) operation in city centres, traffic jams, and for emergency situations, for example, if the vehicle is out of fuel.

Similarly, an emergency mode can be simply programmed into the control algorithms to allow limited vehicle mobility in the event of electrical problems. Should E1 or E2 fail, the transmission can be locked in a state between operation ranges one and two by locking clutches B1 and K2 for low range or between ranges two and three by locking K2 and K3 for high range. Under these conditions, any one of the three motors can propel the vehicle in an absolute emergency.

7.4 Simulations

Simulations were carried out in MathCAD for an SEL-120 installed in a Mercedes-Benz C220 CDI. Using the fuel consumption map, a torque curve was calculated which optimized brake specific fuel consumption. Depending on driver response and transmission mode, the engine is operated in a corridor between and including the maximum power curve and the BSFC-optimized curves. On the right is an efficiency map for an electric motor similar to those used in the SEL-120.

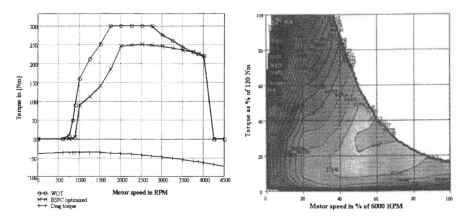

Fig. 7.13 Torque curves for M-B C 220 CDI   Fig. 7.14 Efficiency map for electric motor

In order to maximize system efficiency, the efficiencies of the IC engine, the two electric motors, the battery, and remaining electrical system must be maximized. Depending on the strategy used, the resulting fuel economy in the US city test cycle varies between 4.5 litres/100 km for an extremely economical strategy to 6.5 litres/100 km for a sporty strategy, both compared to the manual transmission baseline of around 6 litres/100 km.

Simulations showed that regenerative braking is responsible for a fuel savings of anywhere from 5 per cent on the motorway to 30 per cent in the city, again depending on control strategy. This figure could be increased through the use of more powerful motors; however, more powerful also means heavier and more expensive, so the achieved gains are a good compromise.

A bonus of parallel hybrid systems is the ability to use the electric motors as a booster, thereby increasing the vehicle's sprint acceleration, as shown in Fig. 7.15.

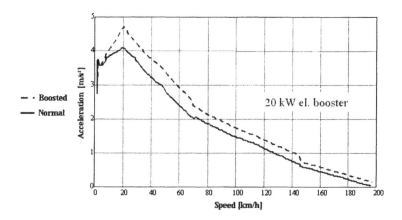

Fig. 7.15 Acceleration curves with and without electric boost

7.5 Conclusion

The advantages of electro-mechanical continuously variable transmissions over their purely electric or mechanical counterparts include higher efficiency, use of tried-and-true technology, and fuel savings of 5–30 per cent from regenerative braking, depending on driving cycle and electric motor capacity. On the negative side are the added cost and weight of the necessary electric/electronic components, due in part to the novelty of this application, and the resulting small scale of production. However, two current trends make the growth of these systems almost inevitable:

(1) energy cell-powered vehicles using electric motors as a primary means of propulsion; and

(2) the threat of mandated reductions in fuel consumption for 'light' trucks in the United States, which has already resulted in hybrid studies in this otherwise conservative sector.

An area of ongoing research is the balance between electric and mechanical in electro-mechanical transmissions. By using more operating ranges, reductions in cost and weight-critical electric/electronic components can be realized, albeit with a cost and weight penalty from greater mechanical complexity. In any case, electrical drive components will be an increasingly common feature of the automotive landscape in the years to come.

7.6 Bibliography

Avery, G., (2000), TU Chemnitz, "Konstruktion und Berechnung eines elektromechanischen Fahrzeuggetriebes", Master's Thesis, June.

Kumura, H., Sugihara, J., Nartita, Y., and **Arakawa, Y.,** Nissan Motor Co., **Nakano, M.** and **Maruyama, N.,** (1999), Jatco Corp., "Development of a dual-cavity half-toroidal CVT", *Proceedings of the International Congress on Continuously Variable Power Transmission*, Eindhoven University of Technology.

Tenberge, P., (1999), TU Chemnitz, "Electric-mechanical hybrid transmission", *Proceedings of the International Congress on Continuously Variable Power Transmission*, Eindhoven University of Technology.

Tenberge, P. and **Hofmann, W.,** (1998), TU Chemnitz, "Mechanisch-elektrische Fahrzeuggetriebe im Vergleich", *VDI Berichte Nr. 1393*, VDI-Verlag.

Tenberge, P. and **Hofmann, W.,** (1999), TU Chemnitz, "Elektromechanisches Hybridgetriebe", *VDI Berichte Nr. 1459*, VDI-Verlag.

Yaegashi, T., (1997), "Toyota hybrid system THS", Toyota Motor Corporation.

(1997), "Toyota readies gasoline/electric hybrid system", *Automotive Engineering International*, SAE International, July.

8

The Design of a Parallel Hybrid Transmission Control System

J Marco, R Ball, and **R P Jones**

Abstract

This chapter will present the design of the transmission control system for the Hybrid Electric Realized Off-road (HERO) vehicle. HERO is a parallel hybrid derivative of a Land Rover Defender. The powertrain consists of a diesel engine and an electrical machine mechanically coupled to the gearbox input shaft. The main function of the transmission control system is to employ the electrical machine to actively assist in gear changes. A systems model of the parallel hybrid transmission was used to analyse and design the transmission control system. The resulting controller is a combination of feedback compensation elements and supervisory rule-bases. The powertrain and transmission control systems were implemented as a simulation model using the Simulink/Stateflow extensions to Matlab. Simulation experiments were used to evaluate the performance of the transmission control system for a variety of off-road and on-road conditions.

8.1 Introduction

Hybrid electric vehicles are widely seen as potential solutions to a variety of current automotive problems, such as the need to reduce exhaust emissions, improve fuel economy, and increase vehicle functionality (Powell, Bailey, and Cikanek, 1996). Hybrid powertrains utilize two propulsion systems, an internal combustion engine, and a battery powered electrical machine, combined together to provide a powertrain in which the two elements can act independently or, as is more usual, complementary to each other.

The work presented in this chapter constitutes part of the overall design activities associated with the HERO programme. HERO is a UK Department of Trade and Industry, DTI, funded, Foresight Vehicle project. HERO is a collaborative venture between the Rover Group, the University of Warwick, and Magnetic Systems Technology. Figure 8.1 presents a schematic representation of the HERO powertrain. The powertrain is made-up of an electrical machine mechanically coupled to the gearbox input shaft. The driver manually controls both gear selection and clutch actuation. The electrical machine has a power rating that provides the vehicle with zero emissions capabilities. The two main reasons why this powertrain architecture was selected are:

- the packaging of the electrical machine within the clutch housing resulted in the minimum amount of modifications being required to the existing mechanical layout and,
- the ability to isolate the internal combustion engine from the electrical machine, via the clutch, enables the vehicle to function as an electric vehicle.

The main disadvantage associated with coupling the machine to the gearbox input shaft is that the increased inertia is detrimental to the driver's ability to change gear (Marco, Ball, Jones, and Lillie, 1999). Because of this, it was decided to design a transmission control system for the vehicle that employed the electrical machine to actively assist in the gear change process. The transmission control system, TCS, is made-up of a low-level feedback control algorithm augmented by a high-level supervisory rule-base. The aim of the feedback algorithm is to control the torque output of the electrical machine, such that it can be employed to equalize the velocity of the input pinion to the velocity of the propulsion shaft reflected through the transmission gearing. The supervisory control system is required to identify when and how the synchronization strategy is to be executed in relation to the current status of the vehicle and the gear change intent of the driver.

Contained within this chapter is some of the technical work associated with the design of the transmission control system for the HERO vehicle. The simulated performance of the transmission control system is presented. The design and simulation platform employed was the Matlab/Simulink/Stateflow tool-set. Simulink is an assignment-orientated non-linear simulation environment (Eriksson and Jacobson, 1999). Stateflow provides a graphical representation of a finite state machine, where states and conditional transitions form the foundation of the system architecture. Combining Simulink and Stateflow enables the integration of continuous and discrete time dynamics with 'mode-switching' type behaviour (Kendall and Jones, 1999).

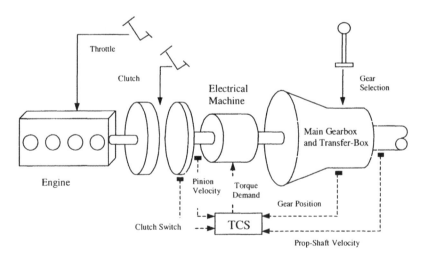

Figure 8.1 Schematic representation of the HERO powertrain

8.2   Hybrid transmission model

Figure 8.2 presents a system level view of the non-linear powertrain model implemented within Simulink. The model contains representative dynamics of the driveline, the vehicle, and the hybrid transmission. The driveline was modelled as a rotational spring and damper in parallel (Ciesla and Jennings, 1995). The vehicle was modelled as a single inertia reflected up-stream of the final drive gearing. The hybrid transmission model, shown in Fig. 8.3, contains representative dynamics of both the electrical machine and the mechanical gearbox. The relationship between the torque demand signal and the machine torque was modelled as a first order lag. The system model of the mechanical gearbox characterizes the dynamics of the compliance within the gear selector and the function of the synchronizer elements associated with each of the forward gears. The model of the synchronizer elements was based on work conducted by (Moir, 1995, Socin and Walters, 1968, and Szadkowski, 1991). The dynamics of the transmission input comprises of the total inertia of the input shaft, clutch plate, and electrical machine. The output shaft of the transmission is made-up of the combined inertia of the gearbox main shaft and the reflected inertia of the transfer-box.

When the transmission is in neutral there is no mechanical link between the transmission input shaft and the output shaft. Therefore, the dynamics of the two inertias are independent of one another. The motion of the input shaft is dependent on the torque losses within the transmission. Conversely, the motion of the main shaft is dependent on the retarding/accelerating torque acting on the vehicle from the external environment.

During synchronization, due to the compliance within the driveline, the motion of the output shaft is no longer directly related to the velocity of the vehicle. The motion of the transmission input shaft, as in neutral, is independent from the remainder of the powertrain. The dynamics of the input shaft are dependent on the torque output from the electrical machine and, therefore, the control algorithms of the transmission control system.

When a gear has been engaged, the input pinion is mechanically coupled to the powertrain. Consequently, during the time interval before the clutch is closed, the dynamics of the input shaft are dependent on both the gearbox loss torque and the retarding/accelerating torque acting on the vehicle reflected through the complete transmission gearing. The motion of the output shaft is assumed to be directly related to the transmission input shaft.

Contained within the powertrain model are two torque losses. First, there is a torque loss due to the external environment, which comprises of rolling resistance, aerodynamic drag and, if the vehicle is traversing an incline, a retarding/accelerating torque due to gravitational effects. Second, there is also a gearbox loss torque that acts on the input shaft of the transmission. Studies into transmission efficiency have identified that the losses within a manual transmission are made-up of two components (Farrall, 1993 and Bartz, 1999). It was found that there are losses due to oil churning and bearing/seal drag and losses due to friction between the gear teeth when a torque is transmitted. For non-zero values of input shaft velocity, the magnitude of the viscous torque loss reduces as the temperature of the lubricant increases. A first-order approximation was employed relating the viscous losses to input shaft velocity for a given lubricant temperature. The second source of torque loss within the manual transmission is due to friction between the gear teeth when a torque is being transmitted. Under such operating conditions, the torque output of the transmission is directly related to the input torque, where the constant of proportionality is a function of the gearing and the efficiency of the transmission.

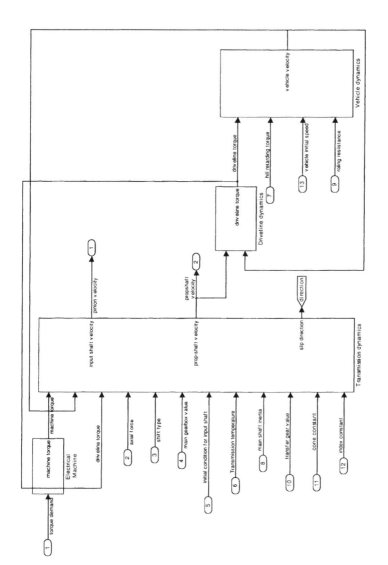

Fig. 8.2 System level representation of the HERO powertrain within Simulink

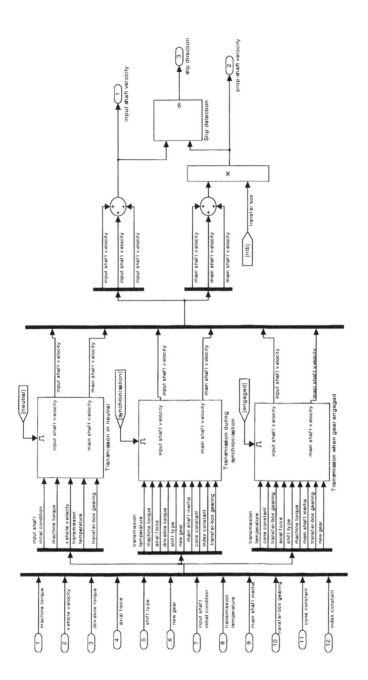

Fig. 8.3 System level representation of the HERO transmission within Simulink

8.3   The transmission control system

Figure 8.4 presents a system level representation of the transmission control system for the HERO vehicle. The control system is made up of a low-level feedback algorithm augmented by a high-level supervisory rule-base. The aim of the feedback compensation algorithm is to control the torque output of the electrical machine, such that it can be employed to equalize the velocity of the input pinion to the velocity of the propulsion shaft reflected through the transmission gearing. The supervisory control system is required to identify when and how the synchronization strategy is to be executed in relation to the current status of the vehicle and the gear change intent of the vehicle operator. The Transmission control system constitutes a single task within the Vehicle Management Unit, the VMU. The VMU is the real-time executive for the HERO vehicle, which schedules the different control functions required within the vehicle. In addition to the TCS, the VMU also controls the operation of the battery management algorithms and the electrical power generation algorithms.

8.4   Design of the feedback control algorithm

8.4.1   The control problem

There were three main elements of the specification for the hybrid transmission:

(i)   the system had to be over/critically damped. Oscillations of the input shaft during synchronization would result in an increased gear change time and an audible noise from the transmission,

(ii)   in order to improve the off-road performance of the vehicle, the hybrid transmission should be able to engage a new gear when the vehicle was travelling up, or down, a steep incline. Such a manoeuvre is known not to be possible within a conventional gearbox due to the rapid rate of acceleration/deceleration of the vehicle, and finally,

(iii)   during normal driving manoeuvres the synchronization time should not exceed 300ms.

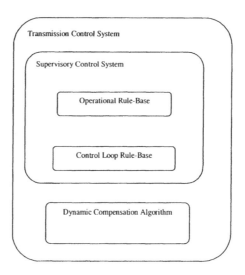

Fig. 8.4 System level view of the transmission control system

Table 8.1 presents the performance specification for the hybrid transmission translated into the frequency domain. Figure 8.5 presents the open loop frequency response of the hybrid transmission. The Bode plot is based on a linear approximation in the form the nominal plant transfer function

$$G_p(s) = \frac{1}{(m_{tc} \cdot J)s^2 + (m_{tc} \cdot B + J)s + B} \tag{8.1}$$

where m_{tc} denotes the time constant of the electrical machine, J denotes the input inertia of the transmission when in neutral, B denotes the total viscous damping coefficient acting on the input shaft and finally, s is the Laplace variable. During the design of the algorithm, the value of B was set to the worst-case operating value for the transmission. Since the phase margin, PM, needed to be greater than 90° the worst case characteristic polynomial represented the minimum damped response of the transmission.

Table 8.1 Performance specification for the controlled hybrid transmission

Control System Performance	Frequency Domain
Dynamic	Phase margin >90^0 Crossover frequency ≅ 25 rads^{-1}
Steady-State	Type II

By comparing the specification against the open loop frequency response, the main conclusions were:

- if the steady-state error was to be reduced to zero under the influence of a ramp-type set-point, then the low frequency gain of the system had to be increased and,
- the PM of the system needed to be increased by approximately 25 degrees.

8.4.2   Control algorithm design

Figure 8.6 presents a block diagram representation of the feedback control system for the hybrid transmission. The method of control system design was the substitution of a continuous algorithm by a discrete equivalent. The design of the control strategy, $G_c(s)$, yielded an algorithm of the form:

$$G_c(s) = \left\{ \frac{k_p \cdot s + k_1}{s} \right\}^2 \cdot \left\{ k_d \frac{T_1 \cdot s + 1}{T_2 \cdot s + 1} \right\} \tag{8.2}$$

where kp denotes the proportional gain, k_I the integral gain, and kd the derivative gain of the lead network. The relative magnitude of T_1 and T_2 yield the frequency location and the amount of phase advance provided by the lead compensator. The two cascaded PI controllers provide the necessary increase in the low frequency gain.

Parameterization of the strategy was an iterative process and involved the graphical interpretation of the system dynamics. Figure 8.7 presents the open loop frequency response of the controlled transmission dynamics. Comparing Figs 8.5 and 8.7, it can be seen that the low frequency gain of the open loop system has been increased by almost 100dB and, in addition, the phase margin of the system has also been increased by approximately 30 degrees. Furthermore, the gradient of the low frequency asymptote indicates that the closed loop system is now of type II.

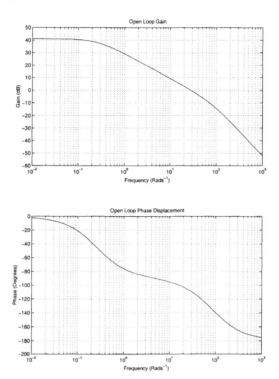

Fig. 8.5 Open loop uncompensated hybrid transmission dynamics

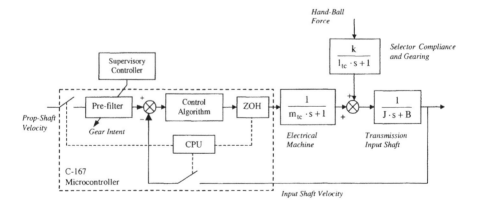

Fig. 8.6 Block diagram of the feedback control algorithm for the transmission control system

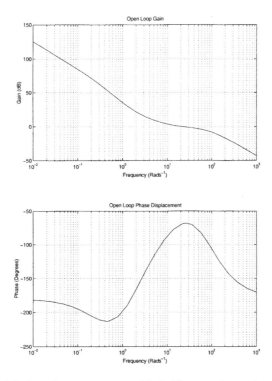

Fig. 8.7 Open loop compensated hybrid transmission dynamics

In order for the control strategy to be implemented within the ECU, the algorithm was converted to its discrete equivalent form, $G_c(z)$, using the pole-zero matching. The design level sample rate was set to 10ms. The digital algorithm, $G_c(z)$, was maintained in its direct form namely, as the ratio of two third order polynomials

$$G_c(z) = \frac{a_0 + a_1 \cdot z^{-1} + a_2 \cdot z^{-2} + a_3 \cdot z^{-3}}{b + b_1 \cdot z^{-1} + b_2 \cdot z^{-2} + b_3 \cdot z^{-3}} \tag{8.3}$$

during the simulation. The term z^{-1} denotes the back shift operator. The numerical coefficients a_n and b_n were stored as floating point constants. Their respective numerical values were rounded to five significant figures. Figure 8.8 presents the feedback control algorithm implemented as a Stateflow model.

The pre-filter shown in Fig. 8.6 comprises of a single gain. The value of gain is scheduled from the supervisory control algorithms in accordance within the gear change intent of the driver, thus ensuring that the correct set-point is applied to the control loop.

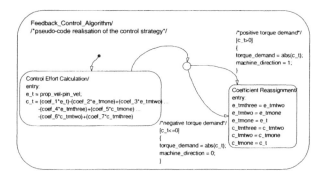

Fig. 8.8 Stateflow realization of the feedback control algorithm

8.5   Design of the supervisory control system

The feedback control algorithm could not be employed in isolation within the HERO vehicle. A supervisory control system was required to integrate the algorithm with the remaining control architecture in order to identify the operational status of the vehicle and the gear change intent of the driver. In order to achieve this functionality, two rule-bases were designed; the operational rule-base and the control loop rule-base.

8.5.1   Design of the operational rule-base

The aim of the operational rule-base was to identify when the driver had placed the hybrid powertrain within the 'gear-change domain'. It thus defines the transition between the TCS and the propulsion-related control algorithms. The gear change domain is defined as the powertrain mode in which the dynamics of the input shaft are isolated from the remainder of the powertrain. In addition, the transfer-box must also be engaged since, as shown in Fig. 8.1, the set-point signal applied to the control loop is derived from a sensor mounted on the propulsion shaft of the vehicle.

When considering the issue of clutch actuation, the studies contained within the literature are mainly concerned with the problem of clutch engagement (Rabeih and Crolla, 1996 and Szadkowski and Moford, 1992). In particular, they deal with the issue of optimally controlling the rate of clutch re-engagement in order to satisfy the contradictory requirements of clutch wear and driveline refinement (Matthews, 1994 and Matthews and Jones, 1993). In the main, the studies assume that the clutch torque capacity increases linearly from zero to unity and is proportional to the pressure acting on the clutch plate. Within an actual vehicle, the driver does not control the value of applied pressure directly, but instead controls the position of the clutch pedal. The point of pedal movement that defines the transition from clutch-locked to clutch-slipping is uncertain and is known to vary during the service life of the vehicle. In order to eliminate the uncertainty associated with the free-play in the actuation mechanism, a binary switch on the underside of the clutch pedal was used to detect when the clutch was fully open.

In order to identify the location of the gear selector, two sensors were integrated within each of the synchronizer elements associated with the five forward gears. One switch is to indicate the gear intent of the driver and the second, to identify when a gear has been engaged. Every possible location of the gear selector is, therefore, identified by a unique code within the TCS.

The gear intent signal is employed by the supervisory control system to identify the desired gear of the vehicle operator and is thus used, as shown in Fig. 8.6, to schedule the gain within the pre-filter. The gear engaged signal indicates to the supervisory control system that a gear has been engaged and thus, when to pass control of the electrical machine back to the propulsion related control algorithms.

8.5.2   Design of the control loop rule-base

The control loop rule-base provides the functionality that is necessary for the proper operation of the feedback control algorithm. The rules are primarily concerned with establishing the required value of gain within the pre-filter in accordance with the location of the gear selector. In addition, the control loop rule-base contains rules that limit the magnitude and rate of rise of the set-point signal.

The desired set-point, des_set_point, to the feedback control algorithm is given by

$$\text{des}\_\text{set}\_\text{point} = \text{prop}\_\text{vel} \cdot n_{tb} \cdot n_1 \tag{8.5}$$

where prop_vel denotes the velocity of the propulsion shaft, n_{tb} the value of transfer-box gearing and n the gear change intent of the driver. The numerical subscript denotes the desired gear. During each sample period the gear intent of the vehicle operator is evaluated. Figure 8.9 presents the rule-base that is employed to ascertain when the vehicle operator has commenced the synchronization process. It should be noted that, in accordance with the physical process of changing gear, the gear signal cannot directly go from neutral to any of the codes relating to a driven gear being engaged. Similarly, for a given gear intent signal the next gear engaged signal is known by the transmission control system.

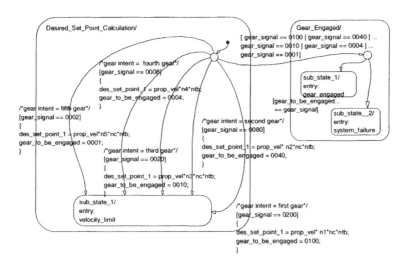

Fig. 8.9 Stateflow realization of the control loop rule-base: set point calculation

After final engagement of the driven gear, if the code contained within the gearbox signal is not the same as that stored within the 'gear_to_be_engaged' parameter, the system assumes the instrumentation has malfunctioned. It should be noted that, the set-point computed within

Fig. 8.9 is only the desired set-point. Before the signal is applied to the feedback system its magnitude and rate of rise are checked to ensure that they are within the desired limits.

Within HERO, the diesel internal combustion engine has a maximum velocity in the order of 400rads^{-1}, whereas the electrical machine has a maximum velocity in excess of 600rads^{-1}. To avoid damaging the clutch, the supervisory control system limits the magnitude of the set-point during synchronization. The value of the upper limit, max_pin_vel, corresponds to the maximum velocity of the internal combustion engine. In addition to limiting the maximum velocity of the transmission input shaft, the minimum velocity of the input pinion is also restricted. This additional functionality reduces the possibility of the driver selecting an inappropriate gear that will result in the engine stalling during re-engagement of the clutch. The lower velocity limit, min_pin_vel, corresponds to a value just above the stall velocity of the engine. The lower velocity limit is only invoked when the vehicle is not stationary. Figure 8.10 presents the rule-base for limiting the velocity of the gearbox input shaft during synchronization.

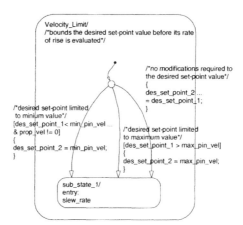

Fig. 8.10 Stateflow realization of the control loop rule-base: velocity limit

Due to high initial values of control effort resulting from the digitization of the control algorithm, a slew rate was required in series with the pre-filter to avoid saturation of the electrical machine. Figure 8.11 presents the structure of the rule-base associated with limiting the rate of rise of the set-point. The term max_grad defines the maximum allowable change in the set-point signal during a single 10ms sample period. The term set-point_tmone, relates to the value of the set-point signal at the previous time interval, namely time-minus–one.

During the initialization phase of the transmission control system, the parameter 'slew_count' is set to unity. The first time the dynamic compensation algorithm is executed, the desired gradient, des_grad, is calculated using the current velocity of the input pinion. Thereon, the value of set-point at the previous time interval is employed in the calculation.

It should be noted that the transmission control system could have been implemented within the simulation environment as a number of interconnected Simulink function blocks, rather than as a finite state machine. There are a number of reasons why Stateflow was employed instead to realize the control algorithms:

- first, it allowed both the dynamic and supervisory parts of the controller to be integrated, holistically, within a single workspace,
- second, the use of Stateflow allowed the data types within the control loop to be specified in terms of those which are to be employed within the actual electronic control module and finally,
- the Stateflow model can be use as an all-exclusive executable specification for the control strategy, which can be passed to the software engineering team when the time comes to code the algorithms within the target hardware (Butts, 1999). Because the Stateflow code is more representative of the final source code program, the possibility of manual coding errors arising from any ambiguity or miscommunication with respect to control functionality is reduced.

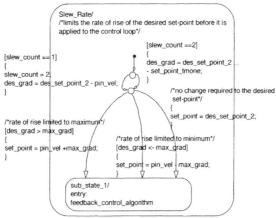

Fig. 8.11 Stateflow realization of the control loop rule-base: slew rate

8.6   Simulated response of the transmission control system

The aim of the simulation study was to verify the functionality of the transmission control system using the non-linear model of the HERO powertrain. The performance of the control system was assessed by means of conducting neutral to drive gear engagement simulations for different gear ratios. A third-order fixed-step integration algorithm was employed, with an integration step size of 1ms. The first simulation study to be conducted was associated with 'standard' gear changes, namely single increment/decrement gear changes through the complete range of the gearbox. The simulation studies were repeated for both an on-road and an off-road environment. Within the confines of the mathematical model, the distinction between on-road and off-road was done on the basis of the coefficient of rolling resistance. Appropriate values for rolling resistance were obtained from (Wong, 1993).

The performance of the hybrid transmission when the vehicle operator attempts a gear change as the vehicle descends a steep incline in an off-road environment is also documented. The angle of the terrain was equal to fifteen degrees, a value which is representative of the level of inclination possible within an off-road environment.

Common to all simulations was the selection of the initial conditions for the model states. For an up-shift, the initial condition for the clutch plate velocity was 300rads⁻¹. Conversely, for a down-shift manoeuvre the same initial condition was altered to 120rads⁻¹. The subsequent state initial conditions are directly dependent on the clutch velocity. All the simulations commenced with the main gearbox in neutral. After a period of 200ms, the gear synchronization procedure was initiated by means of applying a force to the hand-ball of the gear selector. The value of hand-ball force remained constant until the velocity differential between the main shaft and the input pinion reflected to the appropriate gear was reduced to zero. When synchronization was achieved the hand-ball force was set to zero, causing the axial force to decay.

8.7   Discussion of results

Figure 8.12 presents an example of a third to second gear change when the vehicle is traversing level terrain in an on-road environment and when the value of hand-ball force was set to maximum, namely a value equal to 100N.

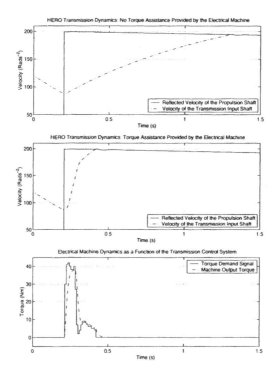

**Fig. 8.12 Third to second, on-road gear change with the maximum
applied hand-ball force**

The figure shows the dynamics of the HERO transmission during a conventional gear change and when the electrical machine is controlled to provide torque assistance. From the figure it can be seen why the transmission control system is required within the HERO vehicle.

Without the torque assistance from the electrical machine the increased inertia on the gearbox input shaft results in an unacceptably long synchronization period. It should be noted that, with the hand-ball force set to maximum, the corresponding synchronization interval represents the minimum realizable value. Any reductions in the value of hand-ball force would result in further increases in the gear change time. When the dynamics of the transmission input shaft are controlled via the TCS, synchronization to the desired gear occurs within the required 300ms. In addition, due to the application of closed-loop control, the gear change time is largely independent of the force applied by the driver.

The second set of simulations investigated the improved off-road functionality of the HERO vehicle. When driving a vehicle up or down an off-road incline, the driver should refrain from interrupting the flow of torque from the engine to the road wheels. If the vehicle operator does attempt a gear change, in particular when the vehicle is accelerating down hill, the velocity profile of the main shaft may be such that gear synchronization is not possible. Figure 8.13 presents the results obtained for a second to first off-road gear change when the vehicle is accelerating down a negative incline of fifteen degrees.

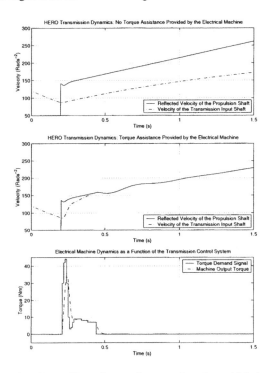

Fig. 8.13 Second to first, off-road gear change when the vehicle is accelerating down a hill

The figure shows both the simulated performance of the transmission control system and the dynamics of the HERO vehicle when the driver must perform the manoeuvre unaided. As it can be seen, even with the value of hand-ball force set to maximum, the driver alone is unable to engage the desired gear. However, within the hybrid transmission that employs torque assistance from the electrical machine, synchronization to the desired gear is achieved. Due to the configuration of the HERO powertrain, if the driver attempts to change gear, there will still be an interruption of torque within the driveline. However, due to the application of closed loop control and because of the nature of the control strategy, drive gear synchronization is now possible when the vehicle is traversing a steep incline in an off-road environment.

8.8 Conclusions

Contained within this chapter is the design of the transmission control system for the HERO prototype vehicle. The transmission control system is hierarchical in nature and comprises of a feedback algorithm augmented by a supervisory control system. The main deliverable from the transmission control system is a manual transmission in which the time required to change gear is now independent of the force applied by the driver. In addition, due to the ability to change gear when the vehicle is accelerating/decelerating a steep incline, the off-road performance of the vehicle is also improved.

8.9 Acknowledgements

The work was supported by EPSRC as part of the Engineering Doctorate programme at the University of Warwick in association with the Rover Group.

8.10 Bibliography

Bailey, K. E. and **Powell, A.**, (1995), A hybrid-electric vehicle powertrain dynamic model, Proceedings of the 1995 *American Control Conference*, Seattle, USA, 167–183.

Bartz, W. J., (1999), Gear oil influences on efficiency of gear and fuel economy of cars, Proceedings of the Institution of Mechanical Engineers, *Journal of Automobile Engineering, Part D,* (**214**), 189–196.

Cielsa, C. R. and **Jennings, M. J.**, (1995), A Modular Approach to Powertrain Modelling and Shift Quality Analysis, *SAE Technical Paper Series*, Number 950149, Society of Automotive Engineers.

Eriksson, A. and **Jacobson, B.**, (1999), Modular modelling and simulation tool for the evaluation of powertrain performance, *International Journal of Vehicle Design,* **21** (2/3), 175–189.

Farrall, S., (1993), *A study into the use of fuzzy logic in the management of an automotive heat engine: electric hybrid vehicle powertrain*, PhD Thesis, University of Warwick.

Kendall, I. R. and **Jones, R. P.**, (1999), An investigation into the use of hardware-in-the-loop simulation testing for automotive electronic control systems, *Control Engineering Practice,* **7**, (11) 1343–1356.

Matthews, J. C. and **Jones, R. P.**, (1993), Optimal feedback control of clutch engagement in an automobile, Proceedings of the 2nd *European Control Conference*, Groningen, 986–991.

Matthews, J. C., (1994*), An Optimization Study on the Control of Clutch Engagement in an Automotive Vehicle*, PhD Thesis, University of Warwick.

Marco, J., Ball R., Jones, R. P., and **Lillie, K.,** (1999), A systems modelling and simulation approach to gear shift effort analysis, Submitted to the *International Journal of Vehicle Design.*

Moir, G. B., (1995), An Investigation into Objective Measures of Gear Shift Quality, Proceedings of the Institution of Mechanical Engineers, Part D, *Journal of Automobile Engineering,* **209,** 273–279.

Powell, B. K., Bailey, K. E., and **Cikanek, S. R.,** (1996), Dynamic modelling of hybrid electric vehicle powertrain systems, *IEEE Control Systems,* **18** (5), 17–33.

Rabieh, E. M. A. and **Crolla, D. A.,** (1996), Intelligent control of clutch judder and shunt phenomena in vehicle drivelines, *International Journal of Vehicle Design,* **17** (3), 319–333.

Socin, R. J. and **Walters, K. L.,** (1968), Manual transmission synchronizers, *SAE Technical Paper Series,* Number 680008, Society of Automotive Engineers.

Szadkowski, A., (1991), Shiftability and shift quality issues in clutch transmission systems, *SAE Technical Paper Series,* Number 912697, Society of Automotive Engineers.

Szadkowski, A., and **Morford, R. B.,** (1992), Clutch engagement simulation: engagement without throttle, *SAE Technical Paper Series,* Number 920766, Society of Automotive Engineers.

Wong, J. Y., (1993), *Theory of Ground Vehicles,* Second Edition, John Wiley and Sons.

Author Index

Subject Index

Printed and bound by CPI Group (UK) Ltd, Croydon, CR0 4YY

16/04/2025

14658831-0001